The Coronary Heart Disease Pandemic in the Twentieth Century
Emergence and Decline in Advanced Countries

The Coronary Heart Disease Pandemic in the Twentieth Century
Emergence and Decline in Advanced Countries

William G. Rothstein
Professor of Sociology Emeritus
University of Maryland Baltimore County
Baltimore, MD
USA

CRC Press
Taylor & Francis Group
Boca Raton London New York

CRC Press is an imprint of the
Taylor & Francis Group, an **informa** business
A SCIENCE PUBLISHERS BOOK

CRC Press
Taylor & Francis Group
6000 Broken Sound Parkway NW, Suite 300
Boca Raton, FL 33487-2742

Printed on acid-free paper
Version Date: 20171003

International Standard Book Number-13: 978-1-138-56950-8 (Hardback)

Library of Congress Cataloging-in-Publication Data

Names: Rothstein, William G., author.
Title: The coronary heart disease pandemic in the twentieth century : emergence and decline in advanced countries / William G. Rothstein.
Description: Boca Raton, FL : Taylor & Francis Group, 2018. | "A science publishers book." | Includes bibliographical references and index.
Identifiers: LCCN 2017043740 | ISBN 9781138569508 (hardback)
Subjects: | MESH: Coronary Disease--epidemiology | Pandemics--history | Coronary Disease--history | History, 20th Century
Classification: LCC RC685.C6 | NLM WG 300 | DDC 616.1/23--dc23
LC record available at https://lccn.loc.gov/2017043740

Visit the Taylor & Francis Web site at
http://www.taylorandfrancis.com

and the CRC Press Web site at
http://www.crcpress.com

Dedicated to
the medical historians of previous generations
who inspired many of us with their collegiality
and commitment to the highest standards of scholarship.

The purely epistemological question – "How do we know" – often diverts attention from more fundamental and (and ultimately political) questions such as: Why do we know this and not that? Why are our interests here and not there? Who gains from knowledge of this and not that?

Robert N. Proctor, *Value-Free Science? Purity and Power in Modern Knowledge* (Cambridge, MA: Harvard University Press, 1991), p. 10.

Contents

List of the Tables

Introduction

Readers who seek a brief summary of the findings, conclusions, and recommendations of this study can read Chapter 12. Chapter 12 does not describe the concepts and methods used in the research, the data used in the analyses, or the nature and scope of the findings.

This study demonstrates that a pandemic of coronary heart disease emerged and declined in most advanced countries in the world during the last two-thirds of the twentieth century. The substantial differences between pandemic coronary heart disease and the disease before and after the pandemic demonstrate that novel causal factors produced the pandemic. The similar characteristics and timings of the pandemic disease in all affected countries demonstrate that it was the result of a single set of causal factors that occurred in all locations at the same times.

Pandemic coronary heart disease differed strikingly from the coronary heart disease early in the twentieth century, which was an uncommon health problem of the elderly. The pandemic emerged in the 1930s and 1940s with rapid increases in adult mortality rates in many advanced countries. Mortality rates increased by much greater amounts for men than women and for older than younger age groups of both sexes in every affected country. The pandemic of this novel form of coronary heart disease reached a peak about mid-century and became a leading cause of death among adults in all affected countries. The peak period ended in the 1970s, after which mortality rates decreased steadily and substantially. The decreases in mortality rates were greater for men than women and for older than younger age groups, which demonstrated that their high mortality rates were caused by the pandemic disease.

Coronary heart disease is a partial or total blockage in the arteries that provide blood to the heart muscle. Lack of an adequate blood supply can produce severe pain, permanently damage the heart muscle, and cause a number of diseases, disability, and death. Early in the twentieth century it was an uncommon disease of the old that was thought to be one of several consequences of "hardening of the arteries." Mortality rates began to increase steadily in the 1930s, notably among persons much too young to experience hardening of the arteries. By midcentury it became the leading cause of adult deaths in many advanced countries on three continents and aroused great concern.

Experts decided that the great increases in coronary heart disease mortality rates in advanced countries were caused primarily by recent dietary and lifestyle changes that resulted from improvements in their standards of living. High rates of the disease would continue indefinitely unless the diets and lifestyles were modified. The experts developed public health programs to implement this theory. They emphasized reduced consumption of types of foods that they believed contributed to the blockages of arteries. They identified high blood pressure and high levels of blood cholesterol as risk factors for the disease. Drugs were developed that decreased levels of these risk factors.

In the 1970s, the high coronary heart disease mortality rates that experts had expected to continue for the foreseeable future began a steady, substantial, and prolonged decrease. The decreases in mortality rates were reversals of the patterns that occurred during the emergence of the pandemic. The decreases occurred at about the same times in all countries, as had the increases. The population groups that experienced the greatest increases in mortality rates during the rise of the pandemic experienced the greatest decreases in mortality rates during its decline. The geographic regions that experienced the greatest increases in mortality rates during the rise of the pandemic experienced the greatest decreases in mortality rates during its decline.

The increases and decreases in coronary heart disease mortality rates in many advanced countries at the same times render it inconceivable that dietary and lifestyle changes were responsible for the pandemic. The advanced countries on three continents that experienced the pandemic are diverse in their economies, cultures, geographies, and social structures. It is beyond the realm of possibility that every one of these countries experienced the same changes in diets and lifestyles at the same times to cause the emergence of the pandemic and then experienced reversals of the changes at the same times to cause the decline of the pandemic. This research will examine the experience of the United States in detail and show that the alleged causal factors did not change in ways or at times that could have made them responsible for either the rise or the decline of the pandemic in that country.

This study uses coronary heart disease mortality rates based on vital statistics in most advanced countries to describe the emergence, peak, and decline of the pandemic. The countries all had detailed and accessible vital statistics for much or all of the period, which greatly reduced concerns over the accuracy of the data. United States government vital statistics were used to describe the increases and decreases in mortality rates over the course of the pandemic for the total population, for age, sex, and race groups, and for geographic regions of the country. Vital statistics for other advanced countries were used to the extent available to describe their patterns of mortality rates over the period of time in this study. The countries included Canada, England and Wales, Australia, New Zealand, and the countries of western and northern Europe. Vital statistics of populous countries in central and south America were examined from 1970 to 2000 to demonstrate that the pandemic did not occur in those countries at that time. The author found that appropriate vital statistics for other countries were either not readily available, not available in sufficient detail, not available in the English language, or not available for an adequate period of time to warrant use in this analysis. The accuracy of mortality reporting in many of these countries was also a concern.

This book has three objectives. One was to develop operational criteria to differentiate the characteristics of a pandemic disease from the normal disease using the multi-national pandemics of influenza, lung cancer, and tuberculosis that occurred in the nineteenth and twentieth centuries. The second was to use the resulting criteria for pandemics to demonstrate that a coronary heart disease pandemic emerged and declined in many advanced countries during the last two-thirds of the twentieth century. The third objective was to evaluate the role of the risk factors that experts have considered to be causes of the high mortality rates during the pandemic. This study did not examine risk factors for coronary heart disease before or after the pandemic. It discussed only primary prevention, which is the prevention of the disease in healthy persons, not secondary prevention, which is concerned with persons who have preexisting heart disease.

It was not feasible for a study of this scope to describe the diagnostic techniques, prevention programs, and therapies for the treatment of coronary heart disease that were used during the pandemic and that occupied so much time and effort of health care professionals and expenditures of health care organizations and governments. The importance of these activities cannot be overemphasized given the prevalence and high mortality rates of the disease. These analyses involve bodies of knowledge and systems of data gathering and analysis that differ from those used in this study. They require analyses of differences among countries. Thousands of articles and many books on these subjects have been published. A detailed history of the diagnosis and treatment of cardiovascular diseases throughout the twentieth century is W. Bruce Fye, *Caring for the Heart: Mayo Clinic and the Rise of Specialization* (New York: Oxford University Press, 2015). A history and analysis of the risk factors for coronary heart disease in the United States is William G. Rothstein, *Public Health and the Risk Factor: A History of an Uneven Medical Revolution* (Rochester, NY: University of Rochester Press, 2003).

This study required the author to transcribe manually more than one thousand numbers from many different sources of widely varying accessibility. Practically all of the numbers were obtained directly or indirectly from publications of United States agencies, the website of the United States Centers for Disease Control, the World Health Organization and its regional offices, and individual governments. Despite the efforts of the author, some unrecognized errors undoubtedly occurred in transcriptions and mathematical computations. The author regrets these and takes full responsibility for them. The reader should also be aware that different sources sometimes used different numbers for the same data points. The author endeavored to identify the most trustworthy sources and used numbers only from them.

The author would like to express his gratitude to Jackie Duffin, Jim Mohr, Shauna Devine, and the late Gerry Grob for their assistance in separate aspects of this research. Each helped the author make better decisions concerning the manuscript. A special debt of gratitude is due to David Grimes, an English physician. His research on coronary heart disease in England in the twentieth century produced findings similar to those of the author. His communications with the author provided many helpful insights that improved the manuscript and strengthened the author's belief in the validity of our similar findings. Last, CRC Press, unlike many other publishers, was willing to publish a manuscript based on evidence that differed from conventional thinking about coronary heart disease. The author is extremely grateful for this decision and the quality of their work preparing the manuscript for publication.

Chapter 1

Pandemics as Historical Events

Pandemics have been observed and feared throughout all of recorded history. This chapter describes characteristics of pandemics of infectious and chronic diseases that will be applied to the coronary heart disease pandemic of the twentieth century. All pandemics occur in large geographic regions and are characterized by relatively rapid increases and decreases in mortality rates for the total population, with greater increases for particular population groups. Many pandemics are the result of novel causal factors that enable them to diffuse quickly throughout the populations. The chapter demonstrates these characteristics of pandemic using the pandemics of influenza, lung cancer, and tuberculosis that occurred in the nineteenth and twentieth centuries.

Throughout history massive outbreaks of severe infectious or chronic diseases have occurred in widespread geographic regions on rare occasions and have been termed pandemics and large scale epidemics. Each pandemic was characterized by relatively rapid increases in mortality rates, a period of peak mortality rates, and similarly rapid decreases in mortality rates. The population groups with the highest mortality rates in each pandemic often differed from the groups that were the typical victims of the disease in normal times. Substantial geographic variations in mortality rates also occurred in each pandemic. Events of this magnitude are so uncommon, distinctive, and extensive in scope that they differ in fundamental ways from the same diseases under normal conditions. Contemporary observers have frequently questioned whether they were the same disease. Pandemics differ from large-scale epidemics by their multinational scope but both share the same basic characteristics.

Pandemics of Infectious and Chronic Diseases

Pandemics and epidemics include both infectious and chronic diseases. Pandemics and epidemics of infectious diseases, which involve pathogens that infect every victim and can be transmitted from person to person, have occurred much more frequently throughout history. As a result, they have been the subject of much more analysis.[1] Pandemics and epidemics of chronic diseases, which have no single causal factor present in every person and cannot be transmitted directly from person to person, became more important causes of death in advanced societies in the twentieth century. Chronic disease pandemics typically occur over longer time periods than infectious disease pandemics.

Modern large scale infectious disease pandemics and epidemics have been carefully studied and analyzed in order to identify them as promptly as possible. Their identification in advanced countries has benefitted from reporting systems that have been steadily refined throughout the twentieth century. Most infectious disease pandemics and epidemics produce increases in the number of cases and visits to physicians, hospitals, and clinics that can be recognized in days or weeks. The patients are often identified promptly because of the characteristics of the disease. This information is transmitted quickly to local public health departments and state and federal agencies, which differentiate the pandemic or epidemic from normal seasonal or other outbreaks of infectious diseases. The agencies carefully monitor its progress and gather relevant statistics from health care providers, often on a daily or weekly basis. They provide information to health care providers and instruct them in improved methods of reporting the disease. They watch for decreases in the number of cases to determine if the pandemic was self-limited. If they believe that the event is likely to recur, they develop methods of gathering periodic information and statistics.

When chronic diseases replaced infectious diseases as the major causes of death and serious illness in advanced countries in the first half of the twentieth century, public health and clinical medicine were confronted with many new challenges. One of them has been the greater difficulty of recognizing pandemics and epidemics of chronic than infectious diseases. Chronic diseases have latency periods of months or years before the occurrence of clinical disease. They often produce symptoms that develop gradually and can be difficult to recognize and interpret. The increase in visits to health care providers usually occurs gradually. It is frequently difficult to determine whether the increase in visits is due to more new cases of the disease or to greater awareness of the disease and its symptoms by persons with the disease. Improved methods of diagnosis can detect unrecognized cases of a chronic disease and these can be mistaken for increases in the number of new cases. Modifications of the methods of diagnosis of a chronic disease often produce increases in the number of new cases of the disease. Measuring the decline of a chronic disease pandemic or epidemic can be equally problematic. Diagnoses at earlier stages of the disease that occur with improved awareness and knowledge produce longer lengths of survival of persons with the disease. Some of these problems also occur with infectious diseases, but they are much more frequent and difficult to solve for chronic diseases.

Another problem with pandemics of chronic diseases has been the difficulty of differentiating the pandemic from the steady increases in the prevalence of many chronic diseases in advanced countries since the early twentieth century. This has resulted from the increased numbers of elderly persons, who are the most frequent victims of chronic diseases. It is clearly difficult to differentiate an emerging pandemic or epidemic of a chronic disease from an increase in the number of cases of a chronic disease produced by more elderly persons in the population. Many chronic disease pandemics and epidemics are identified only when their mortality rates are decreasing, so that the decreases in mortality rates can be an essential factor in recognizing a pandemic.

Despite these difficulties, methods of reporting diseases in the twentieth century have advanced so much that outbreaks of chronic diseases that are later recognized as pandemics or epidemics are often detected in several years. The atypical characteristics of the patients in many of these diseases usually alert health care providers to the novel situation. For example, articles in the 1920s and 1930s in medical journals in the United States and other advanced countries commented on the increase in coronary heart disease cases. The articles expressed confusion about the nature of the disease and surprise at the young ages of some patients. To take another pandemic disease, physicians in the 1930s became concerned with the presence of lung cancer in patients who were never exposed to the known carcinogens. Statistical evidence of increases in lung cancer mortality rates was publicized in the United States in the 1930s and led to much speculation as to the reasons for its occurrence.[2]

Individual pandemic and epidemic diseases have differed in their previous histories in the affected countries. Some diseases were rare or unknown or considered foreign and exotic. Their emergence is usually recognized promptly, although an understanding of the disease may take many years. Other diseases were known to the medical and public health communities. Their emergence is typically marked by difficulty in differentiating the pandemic from the normal occurrence of the disease.

Pandemics and epidemics can attack populations in both poor and good health. Many pandemics developed during periods of upheavals in social and hygienic conditions, including wars, depressions, and industrialization. These indicate that deteriorating general population health was a causal factor. For example, during World War I the reduction in the standard of living in some European countries produced a resurgence of tuberculosis, a pandemic disease that had been steadily decreasing in its mortality rates. Conversely, major pandemics occurred in the mid-twentieth century in advanced countries that were experiencing impressive improvements in their health and living conditions. These included the paralytic polio, lung cancer, AIDS, and coronary heart disease pandemics. It can therefore be concluded that pandemics are highly atypical events that are often unrelated to the overall health of the population.[3]

Pandemics and large-scale epidemics are always the result of novel changes in social conditions that affect large populations in widespread geographic areas. A pandemic can develop only if extraordinary social conditions enable the causal factors to diffuse throughout very large populations and geographic areas. The extremely rarity of rapid and widespread diffusions is demonstrated by the innumerable localized epidemics that have attacked small population groups in restricted geographic areas for short periods of time and then receded. Some of them did not become pandemics or large scale epidemics because they lacked the distinctive social conditions that would have enabled them to diffuse quickly among large geographic areas.[4]

The reasons for the decline of pandemics that occurred in multiple countries with hundreds of millions of residents are varied. Many pandemics declined spontaneously for unknown reasons, just as they emerged for unknown reasons. Others declined because of knowledge of the social conditions that enabled the causes of the pandemic to spread among the population. An example is the development of massive programs to discourage cigarette smoking and restrict the sale of cigarettes in the second half of the twentieth century. Prevention of infectious disease pandemics with vaccines, such as the polio pandemic in advanced countries, has contributed both by protecting individuals directly and exposing healthy persons to fewer sick persons. Medical treatment of patients who develop the disease has rarely been a significant factor in the decline of multinational pandemics. Useful treatments were not available for many pandemic diseases and in other pandemics the numbers of persons with the disease overwhelmed the resources of the health care system.

Criteria for Pandemics

The criteria for pandemics developed in this paper are based on three multinational disease outbreaks that are widely accepted as pandemics and have been the subject of much research: influenza and lung cancer in the twentieth century and tuberculosis in the nineteenth and early twentieth centuries. Influenza was a world-wide infectious disease pandemic that occurred in three waves in less than three years. Tuberculosis, an infectious disease, and lung cancer, a chronic disease, produced pandemics in advanced countries that took a number of decades to emerge, reach a peak, and decline, although the lung cancer pandemic had not ended by the early twenty-first century. All three were highly atypical events characterized by relatively rapid increases in mortality and morbidity rates, a period of peak rates, and then relatively rapid decreases in the rates.

Most pandemics have causal factors that differ from the causes of the disease in normal times. It is often difficult to identify these novel causal factors, but their presence can be recognized by the population groups and geographic areas that are most affected. All diseases affect specific population groups and geographic areas to different extents and this can often be used to differentiate a pandemic from the normal form of the disease. It will be shown that the most severely affected population groups in the three pandemics differed from the groups that usually develop the disease in normal times. After its emergence in particular atypical population groups, the pandemics diffused to other population groups, but the groups affected earliest continued to have the highest mortality rates. The decline of the pandemics was the reverse of its emergence for each affected population group. The greatest decreases in mortality rates occurred among the atypical groups that had the highest mortality rates at the peak of the pandemic. This has often been used as evidence that the pandemic was declining.

Geographic regions experienced variations in pandemic patterns similar to population groups. Most pandemics emerged in particular geographic locations and then spread to other locations. The decline of the pandemics produced greater decreases in the mortality rates of the geographic areas with the highest rates of the disease at the peak of the pandemic.

Evidence that a pandemic disease outbreak is not part of a normal pattern requires comparison with secular trends in overall mortality rates. This analysis will demonstrate that the increases and decreases in the mortality rates of the pandemics were not part of secular trends in mortality rates from all other causes.

Influenza

The great influenza pandemic of 1917-20 clearly met the criteria for a major pandemic in terms of the rapid increase in mortality rates at the onset of the pandemic, the extremely high mortality rates at its peak, the rapid decrease in mortality rates as it declined, and its highly atypical victims. It appeared first in early 1917, intensified quickly and killed tens of millions of persons in rich and poor countries around the world in three waves in the next two years, and then declined with equal rapidity.

Probably the most distinctive feature of the influenza pandemic was that many of its victims were young adults, who were extremely unlikely to die of the disease in normal times. According to Crosby, "it killed an unprecedentedly large proportion of the members of a group who, according to records before and since, should have survived it with no permanent injury."[5] As the pandemic declined, the mortality rates of young adults decreased more than those of other groups and the highest rates resumed among the groups that had been traditional victims of the disease.

Influenza is normally a mild infectious disease, but it can lead to pneumonia that is responsible for most deaths. The pandemic influenza virus differed from the normal influenza virus in several respects. It infected many more persons than other strains, possibly because of its greater ability to be transmitted from person to person, and produced a much more frequent occurrence of severe pneumonia. One hypothesis, developed decades after the pandemic, is that the lung damage was caused not by the infection itself but by the overreaction of the immune system to the infection. The lungs were filled with fluid and cells so that oxygen could not be absorbed into the blood stream. Although this pandemic was highly atypical, so many epidemics and pandemics of influenza with varying characteristics have been reported for more than two thousand years that it is not possible to say that it was unique.[6]

The influenza pandemic consisted of three waves of short duration: a first wave with low mortality rates in the spring of 1917 during World War I in both the United States and Europe, an extremely severe second world-wide wave in late 1918 as the war was ending, and a third wave in early 1920 after the war had ended. The second and third waves in the United States were large enough to produce substantial increases in monthly total mortality rates in all of the states that were in the death registration area. The first wave in early 1917 produced no increase in total monthly crude mortality rates in the death registration area. The total mortality rates per 1000 population in 1917 were 15-17 from January to April, which were similar to the rates in the same months of 1916. The second wave of the pandemic in late 1918 produced astonishingly high total mortality rates per 1000 population of 44 in October, 25 in November, and 22 in December. These may be compared to rates of 12-14 in the same months of 1917. The rates then decreased steadily in 1919 to 10-12 from May through December. The third wave in early 1920 produced a peak total mortality rate per 1000 population of 24 in February, after which mortality rates decreased to about 10-12 after April.[7]

The pandemic spread quickly and widely among geographic areas because of the social conditions produced by World War I. One of the first outbreaks of the disease in the United States occurred on a military base in rural Kansas in early 1918 even though influenza is usually an urban disease. The disease most likely was brought there by some newly drafted soldier or soldiers who contracted the disease in their community or communities. Rural communities are most likely because the influenza virus can be directly transmitted from ducks and other birds to humans and these fowl are more likely to be found on farms. Had these soldiers lived in one of the many isolated farming communities and not been drafted, the pandemic might have been confined to a local outbreak.[8]

The spread of the first wave of the pandemic was exacerbated by the hundreds of thousands of recently drafted American soldiers from many locations who were sent to poorly constructed military camps designed to house a small fraction of their numbers. Dehner observed about the first and second waves: "The situation–young men from diverse areas (countryside, town, and city) crowded together, under the stress of training, and connected to various regions through a steady stream of transfers–was perfect for incubating and spreading a new infection, especially a respiratory infection." Many patients on army bases were not hospitalized but housed in the barracks, where they spread the disease to others. The soldiers who were sent to the camps were often the forerunners of the disease in the communities near the camps. According to Crosby, "time and again, the first news of local influenza in a city's newspapers referred to army camps and army personnel."[9]

A much more severe second wave of the influenza pandemic occurred in the fall and winter of 1918. As hundreds of thousands of soldiers were transported around the country by ship and train, according to Dehner, "the virus began to hopscotch along the nation's transportation routes" and attack the civilian population. The soldiers spread the disease to ship and railroad employees as they traveled so that "merchant seamen and railroad men must have been the vanguard of the pandemic in hundreds of cities and towns," according to Crosby. The transmission of infection from person to person was exacerbated by the inability of hospitals to provide sufficient beds for the victims so that many cases remained in the community. Physicians and nurses were in short supply in communities because of their service in the war effort. After the armistice in November 1918, soldiers returning from Europe to locations throughout the United States continued to spread the disease.[10]

International travel related to the war effort provided the opportunity for the second wave to traverse the world in four months and infect hundreds of millions of persons. During 1918 1.5 million American soldiers traveled to Europe by ship across the Atlantic Ocean and many brought the disease with them. They spread influenza in Europe to soldiers and civilians in many countries. The crews of the navy ships on which the soldiers sailed became infected and spread influenza to the residents of the port cities and the crews of commercial ships in Europe and the United States. The commercial ships traveled to locations on every continent, bringing the disease with them in the summer of 1918.[11]

The personal characteristics of the victims of the influenza pandemic of 1918-20 differed in striking ways from the normal victims of the disease. The most frequent victims of this extremely common disease in normal times are infants, young children, and the elderly. Those in poor health are much more susceptible to the disease than those in good health.

An investigation of the personal characteristics of Americans who died from influenza during the pandemic requires the use of non-governmental sources because government mortality statistics were incomplete and unreliable. The most useful mortality statistics for the influenza pandemic in the United States are included in an analysis of the mortality rates of more than 10 million low income urban life insurance policyholders of the Metropolitan Life Insurance Company. This major research project will be described in detail in Chapter 3. The mortality rates from influenza and pneumonia are combined because the two causes of death were not separated at that time. Mortality rates from pneumonia were relatively stable over time so that any short-term changes were due to the effects of the influenza pandemic.[12]

Young adults, who have low influenza mortality rates in normal times, were the Metropolitan Life Insurance Company policyholders who experienced the greatest increases in influenza and pneumonia mortality rates. Comparing annualized average influenza and pneumonia mortality rates per 1000 for Metropolitan Life Insurance Company industrial policyholders in 1911-17 to their rates in 1918, the rates for white men ages 25-34 increased from 0.8 to 12.0, for colored (the term used by the authors) men from 1.6 to 7.6, for white women from 0.4 to 9.0, and for colored women from 0.8 to 6.3.[13]

The experiences of the young men in training on military bases who were the first victims of the pandemic provide additional information about these victims. These young and recently recruited soldiers were in excellent health. They were very likely to be infected because of their close proximity to other soldiers on the bases, but their good health should have enabled practically all of them to recover. However, many American physicians observed that a large proportion of the soldiers and other young victims who died were in good health. At the December 1918 meeting of the American Public Health Conference, it was agreed that many victims "had been in the best of physical condition and freest from previous disease." One hypothesis for the high death rates of this age group is the effectiveness of their immune systems, which produced the strongest reaction to the infection and the most damage to their lungs.[14]

Considering other Metropolitan Life Insurance Company population groups, the elderly, who typically have the highest influenza and pneumonia mortality rates, had much smaller increases in their mortality rates than young adults. For those ages 65-74, annualized average mortality rates per 1000 Metropolitan Life Insurance Company industrial policyholders increased from 1911-17 to 1918 for white men from 8.4 to 10.3, for colored men from 9.5 to 11.5, for white women from 8.4 to 9.3, and for colored women from 9.6 to 11.3.[15]

These highly atypical patterns of large increases in mortality rates among young adults and small increases among the elderly occurred throughout the world. They had a major impact on European armies on both sides of the conflict. It was observed in both advanced and developing countries.[16]

In normal times, colored Metropolitan Life Insurance Company policyholders had higher mortality rates from influenza and pneumonia than their white counterparts, but during the pandemic white men and women had higher mortality rates than colored men and women. From October 1917 to September 1918, before the severe second wave of the pandemic, influenza and pneumonia crude mortality rates per 1000 for all ages were 3.2 for colored men, 1.6 for white men, 2.2 for colored women, and 1.2 for white women. During the second wave of the pandemic, from October to December 1918, crude mortality rates increased by an extraordinary amount for all groups and produced higher mortality rates among white men and women than their colored counterparts. The mortality rates per 1000 were 18.4 for white men, 15.2 for colored men, 17.2 for white women, and 15.0 for colored women. The decline of the second wave restored the normal pattern. From April to June 1919, crude mortality rates per 1000 were 1.8 for colored men, 1.0 for white men, 1.7 for colored women, and 1.0 for white women.[17]

The pandemic exhibited a different pattern among young children, that of an exacerbation of the normal pattern of influenza and pneumonia. Mortality rates increased by large amounts for all groups, with colored children having higher mortality rates than white children both before and during the pandemic. Among those ages 1-4, annualized average mortality rates per 1000 from influenza and pneumonia increased from 1911-71 to 1918 for white boys from 2.4 to 7.3, for white girls from 2.3 to 7.7, for colored boys from 5.5 to 12.3, and for colored girls from 5.1 to 14.0.[18]

Pandemics also differ from normal diseases because they often emerge in geographic locations that are not typical hosts to the disease. The 1918-19 influenza pandemic in the United States began among young and exceptionally healthy soldiers on military bases in rural areas. It then spread to urban areas, the opposite of the pattern of most infectious disease pandemics and epidemics.

Infectious disease pandemics are also distinguished by their rapid diffusion among large geographic areas. The first wave of the 1918 influenza pandemic spread from the United States to Europe in two months, brought by American troops sent to Europe during World War I. The second wave spread around the world in four months.[19]

The different causal factors of the influenza pandemic compared to normal influenza are also indicated by the different characteristics of the persons who developed the disease during and after the pandemic. The atypical persons with the highest influenza rates during the pandemic experienced the greatest decreases in their mortality rates as the pandemic declined, restoring the usual situation. When influenza mortality levels returned to their normal levels in the United States in the 1920s, the normal pattern resumed of higher mortality rates among infants and children, the elderly, and the colored population groups of Metropolitan Life Insurance Company policyholders. Sex differences in influenza mortality rates were small throughout the pandemic.[20]

Lung Cancer

The lung cancer pandemic that began in the mid-twentieth century is an example of a pandemic of a noninfectious disease. Lung cancer before the second quarter of the twentieth century was an extremely rare disease that was almost always caused by exposure to particular working conditions. It occurred at that time among men who worked in industries that exposed them to certain chemicals, minerals, or gases. The major industries were not located in cities and were not increasing in employment.[21]

Lung cancer became a pandemic because new technologies produced novel causal factors that had their greatest impact on persons who were not the typical victims of the disease. The technologies consisted of new methods of manufacturing inexpensive cigarettes and advertising them widely. Cigarette smoking then became popular among population groups that were not at previous risk of lung cancer. Lung cancer was not the only disease affected by the increase in cigarette smoking, but it was the most striking because of the great increases in mortality rates.[22]

The lung cancer pandemic began in the United States with increased lung cancer rates in white men who had never worked in the small number of industries where the disease was known to occur. The increase was surprising because employment in those industries had not increased prior to the increase in lung cancer mortality rates. In addition, some of the highest lung cancer rates early in the pandemic occurred in urban areas that did not have the industries involved.[23] The impact of the pandemic on white men during its early years is shown by the large differences in lung cancer mortality rates among population groups. In the United States in 1940 the mortality rates of cancers of the respiratory system per 1000 at ages 55-64 were 0.5 for white men, 0.2 for nonwhite men, 0.1 for white women, and 0.1 for nonwhite women. By 1960 smoking had become more widespread among black men as shown by the large increase in their lung cancer mortality rates. The mortality rates in 1960 were 1.5 for white men, 1.5 for nonwhite men, 0.2 for white women, and 0.2 for nonwhite women.[24]

The higher mortality rates for men than women helped disprove the early belief that the increase in lung cancer mortality rates was caused by exposure to new factors in the environment, such as air pollution from motor vehicles and other sources. Population groups that were equally exposed to the environmental factors varied greatly in their lung cancer mortality rates. It also disproved the theory that the increase was caused by improved diagnosis through greater use of x-rays, which would have affected men and women equally.

It was gradually recognized that a temporal association existed between the popularization of cigarette smoking in the 1930s and the increases in lung cancer rates in the 1940s in advanced countries. Newly invented machinery mass produced inexpensive cigarettes and the newly invented motion pictures and radios helped popularize their use in the 1920s and 1930s. World Wars I and II also helped popularize cigarette smoking among soldiers. Cigarette sales per capita based on federal cigarette tax collections reached their peak from 1960 to 1980. The steady increase in cigarette sales over such a long period was partly due to the popularization of smoking among new groups of smokers, primarily women. Annual cigarette sales per person were 977 in 1930, 1349 in 1940, 2390 in 1950, 2645 in 1960, 2534 in 1970, 2752 in 1980, 2060 in 1990, 1551 in 2000, and 1001 in 2010.[25]

The peak and decline of the pandemic was characterized by changes in the lung cancer mortality rates of specific groups. As it reached its peak, high mortality rates expanded from men to include women. As it declined, mortality rates decreased for all groups. Mortality rates per 1000 from cancers of the trachea, bronchus, and lung increased for white men ages 55-64 from 0.9 in 1950 to 1.9 in 1970 and 2.1 in 1990 and decreased to 1.5 in 2000 and 1.0 in 2010. The rates for white women ages 55-64 increased from 0.1 in 1950 to 0.4 in 1970 and 1.1 in 1990, and decreased to 1.0 in 2000 and 0.7 in 2010. The rates for black men increased from 0.7 in 1950 to 2.3 in 1970 and 3.6 in 1990, and decreased to 2.2 in 2000 and 1.6 in 2010. Those for black women increased from 0.1 in 1950 to 0.3 in 1970 and 1.2 in 1990 and decreased to 1.0 in 2000 and 0.7 in 2010. [26]

Evidence of a causal association between cigarette smoking and lung cancer was provided in community surveys of smokers and nonsmokers. These studies used statistical comparisons of mortality rates of groups of smokers and nonsmokers because many smokers did not develop lung cancer during the period of the studies and some nonsmokers did develop lung cancer. The most rigorous studies selected large groups of healthy smokers and nonsmokers who were similar in other characteristics and followed them for many years. The studies consistently found much higher lung cancer mortality rates among the smokers. The studies also found a substantial dose-response relationship: the more cigarettes smoked per day or the greater the number of years when persons smoked, the higher the probability of developing lung cancer. In addition, smokers who stopped smoking had lower lung cancer rates than those who did not stop smoking. Lengthy studies in the community were needed because of the long latency period required for cigarette smoking to cause lung cancer. The early statistical studies were extremely controversial because many medical scientists refused to accept causal relationships based on statistical analyses of mortality rates in community groups. Ultimately, the consistency and strength of the association between smoking and lung cancer and the strong dose-response relationship led to its widespread acceptance.[27]

Pandemics and epidemics of noninfectious diseases such as lung cancer that result from new causal factors produced by new technologies often decline after government interventions that restrict the availability of the etiological factors, mandate safety devices or other interventions, and educate the public. The primary focus has been on prevention using social programs, not new therapies to improve treatment of the diseases. The lung cancer pandemic in advanced countries produced programs designed to reduce cigarette smoking rates through education and restrictive legislation. These programs reduced rates of cigarette smoking substantially, although it continued to be the single most important preventable cause of morbidity and mortality in advanced countries in the early twenty-first century. Comparable pandemics and epidemics have declined because of obsolescence of the technologies or their replacement by safer technologies.

Tuberculosis

Tuberculosis was the most widespread and deadly pandemic of long duration in all advanced countries during the nineteenth and early twentieth centuries. Godias Drolet, who in 1947 wrote the most thorough quantitative analysis of this period of tuberculosis, observed:

"The epidemic nature of tuberculosis is undoubted, though in contrast with the more acute infections it manifests itself in comparatively slow-moving cycles, sometimes across several generations. . . . But it still follows the same general course: widespread at times; common to particular localities or among certain groups; moves wavelike with definite rises when it comes across virgin soil and gradually declines as resistance is evolved; finally, it may flare up again where contact has been lost."[28]

Evidence that a pandemic of tuberculosis emerged in the early nineteenth century is indicated by mortality rates in major cities at that time. Deaths from tuberculosis in the nineteenth century were usually reported accurately because of the distinctive symptoms of the late stages of the disease, the familiarity of physicians with it, and the younger ages of most victims that made comorbidities uncommon. Urban mortality rates are more trustworthy because the causes of rural deaths were rarely reported at that time. A study of tuberculosis mortality in London found that the disease was responsible for more than 20 percent of all deaths in 1655, which decreased to 13 percent in 1715, and then increased to 30 percent in 1801. In New York City in 1804 tuberculosis was responsible for 23 percent of all deaths.[29]

The pandemic reached its peak in different countries at different times in the second half of the nineteenth century, after which mortality rates decreased gradually but steadily. In England and the state of Massachusetts in the United States, the pandemic peaked at midcentury and then declined. In England annualized age standardized average tuberculosis mortality rates per 1000 decreased from 4.6 in 1851-5 to 4.2 in 1861-65, 3.7 in 1871-75, 3.2 in 1881-85, 2.7 in 1891-95, and 2.1 in 1901-05. In Massachusetts, crude mortality rates for tuberculosis of the respiratory system per 1000 decreased from 3.7 in 1861 to 3.4 in 1870, 3.1 in 1880, 2.6 in 1890, 1.8 in 1900, and

1.4 in 1910. Most countries in continental Europe, including France, Norway, Sweden, Germany, Switzerland, Austria, and Hungary, experienced the beginning of their decreases in tuberculosis mortality rates somewhat later in the nineteenth century.[30]

The pandemic came to an end in the early twentieth century. In England, age-standardized annualized average mortality rates per 1000 from tuberculosis decreased from 2.7 in 1891-95 to 2.1 in 1901-5, 1.2 in 1921-25, 0.7 in 1941-45, and 0.2 in 1951-55. In Massachusetts the crude tuberculosis mortality rate per 1000 of 1.4 in 1910 decreased to 1.0 in 1920, 0.3 in 1940, and 0.2 in 1950.[31]

The tuberculosis pandemic in advanced countries was not caused by broad long-term social changes, even though they are usually used to explain it. These are urbanization and industrialization and the resulting unhygienic living and working conditions of most of the population. The pandemic emerged decades before these conditions developed on any substantial scale. Furthermore, the pandemic began to decline in England and Massachusetts in the second half of the nineteenth century, decades before improvements occurred in the health related living conditions of the population.

The lack of improvement in the healthiness of living conditions during the second half of the nineteenth century is indicated by trends in infant deaths in the first year of life, which are considered to be the best measures of living conditions in the nineteenth century. Most infant deaths can be prevented by better housing, elementary public health programs, improved sanitation, potable water supplies, milk that is safe to drink, and similar measures. Living conditions did not become more hygienic in England and Massachusetts in the second half of the nineteenth century as indicated by the lack of any real decrease in infant mortality rates. In England, annualized average male infant mortality rates per 1000 were 172 in 1851-55, 166 in 1861-65, 167 in 1871-75, 152 in 1881-85, 165 in 1891-95, and 151 in 1901-05. Female infant mortality rates in the same years were 141, 136, 138, 125, 135, and 124. In Massachusetts, the annualized average infant mortality rates were 131 in 1851-54, 143 in 1860-64, 170 in 1870-74, 161 in 1880-84, 163 in 1890-94, and 141 in 1900-04.[32]

The victims of tuberculosis during the pandemic did not develop the disease shortly after being infected with the bacillus. Throughout the pandemic autopsies and tuberculin tests in Europe and the United States showed that the great majority of the urban adult populations of the affected countries had been infected at young ages with the bacillus.[33] Yet in 1914, white males and females and black males ages 15-24 had lower tuberculosis mortality rates than in any older age group up to age 55-64 (Table 1.1).

Tuberculosis mortality rates varied considerably among groups that had similar rates of infection with the bacillus. In the United States, the native born white and black groups and the foreign born nationality groups had similarly high rates of infection with the tubercle bacillus, However, the groups varied greatly in their tuberculosis mortality rates and some low socioeconomic immigrant groups had very low mortality rates.[34]

These striking characteristics of the disease raised questions about the role of the tuberculosis bacillus during the pandemic. One American physician observed in the early twentieth century: "The tubercle bacillus, although being the 'sine qua non' of tuberculosis, is after all practically, especially from a prophylactic or hygiene point of view, a minor factor in its multitudinous etiological factors."

The proportion of the urban populations in all advanced countries who were infected with the bacillus in childhood remained high early in the twentieth century, even though tuberculosis mortality and morbidity rates decreased steadily. In London in 1930-31, when tuberculosis mortality rates were a small fraction of the rates at the peak of the pandemic in the mid-nineteenth century, a sample of 1003 children ages 14-15 found that 82 percent tested positive for tuberculosis using tuberculin tests. In the United States similar tests in children of the same ages at about the same time found positive test results in 42 percent of 8045 children in New York City, 31 percent of 1000 children in Chicago, 80 percent of 2678 children in Philadelphia, 70 percent of 2045 children in Minneapolis, 47 percent of 3500 children in San Francisco, and 17 percent of 2467 children in Detroit.[35]

Tuberculosis mortality rates varied considerably among geographic regions, another basic characteristic of pandemics. They were much higher in urban than rural areas. Tuberculosis mortality rates in Europe were highest in the countries of northern and central Europe, which were the most urbanized and wealthiest. Climate was not a key factor because rural areas in all countries had lower mortality rates than urban areas.[36]

The population groups with the highest rates of tuberculosis during the pandemic in the United States differed significantly from those that develop tuberculosis in normal times. One difference was the higher tuberculosis mortality rates of nonwhite men and women than their white counterparts. Tuberculosis mortality rates in the United States are available for age and sex groups of the white and nonwhite populations as the pandemic declined from 1914 to 1950, although not for earlier periods (Table 1.1). The data are for the states enrolled in the Death Registration Area, which included 23 states in 1914, 47 states in 1930, and all 48 states in 1950. In 1914, during the pandemic, nonwhite men and women had considerably higher tuberculosis mortality rates than their white counterparts. As the pandemic declined, the rates of nonwhite men and women decreased more than the rates of white men and women. This indicates that the 1914 differences were a characteristic of the pandemic, not the normal pattern of the disease. At ages 35-44 the tuberculosis mortality rate per 1000 in 1914 of nonwhite men was 2.9 more deaths than white men, which decreased in 1950 to 0.8 more deaths. The rate for nonwhite women ages 35-44 in 1914 was 2.5 more deaths than white women, which decreased to 0.6 more deaths in 1950. At ages 55-64, nonwhite men had a tuberculosis mortality rate per 1000 in 1914 that was 3.1 deaths more than white men, but the difference dropped to 1.0 deaths in 1950. At the same ages, the tuberculosis mortality rate per 1000 for nonwhite women was 1.4 more deaths than white women in 1914 but only 0.4 more deaths in 1950.

The higher mortality rates and greater decreases in mortality rates of the nonwhite than the white population early in the century were the opposite of what would be expected by their geographic locations. Tuberculosis mortality rates were higher in urban than rural areas, but in 1914 a much larger proportion of blacks than whites lived in rural areas. In 1910 72 percent of the black population lived in rural areas compared to 51 percent of the white population. The black population became more urban early the century as their tuberculosis mortality rates decreased, but higher rates of urbanization should have produced continuing high mortality rates.[37]

Table 1.1 United States Tuberculosis Mortality Rates by Age, Sex, and Race, 1914-1950
(Rates per 1,000 persons)

Age	*White*		*Nonwhite*	
	Male	*Female*	*Male*	*Female*
15-24				
1914	1.2	1.3	4.9	5.5
1930	0.4	0.7	2.4	3.0
1950	0.1	0.1	0.4	0.6
25-34				
1914	2.0	1.7	5.2	5.0
1930	0.8	0.8	3.2	3.0
1950	0.1	0.1	0.8	0.8
35-44				
1914	2.3	1.4	5.2	3.9
1930	0.9	0.6	2.8	2.1
1950	0.3	0.1	1.1	0.7
45-54				
1914	2.2	1.1	4.5	3.3
1930	1.1	0.5	2.4	1.7
1950	0.5	0.1	1.5	0.6
55-64				
1914	2.2	1.3	5.3	2.7
1930	1.2	0.6	2.0	1.5
1950	0.7	0.2	1.7	0.6

Source: Anthony M. Lowell, Lydia B. Edwards, and Carroll E. Palmer, *Tuberculosis* (Cambridge, MA: Harvard University Press, 1969), p. 69.

Sex differences in tuberculosis mortality rates occurred during the pandemic and these decreased as the pandemic declined, which indicates that they were a characteristic of the pandemic (Table 1.1). For example, mortality rates per 1000 persons for white men ages 35-44 were 0.9 greater than those for white women in 1914 but only 0.2 greater in 1950. Among those ages 55-64, white men had 0.9 more deaths per 1000 persons in 1914 but only 0.5 more in 1950. Nonwhite men ages 35-44 had 1.3 more deaths per 1000 persons than nonwhite women in 1914 but only 0.4 more deaths in 1950. Nonwhite men ages 55-64 had 2.6 more deaths per 1000 persons in 1914 than nonwhite women, but only 1.1 more deaths in 1950.

Public health programs in the United States to control tuberculosis and improve living conditions in the early twentieth century prevented many cases of tuberculosis but they were adopted decades after tuberculosis morbidity and mortality rates started to decrease. The programs were established almost solely in urban areas but the decreases occurred in both urban and rural areas. Thus the programs could not explain the greater decreases in the mortality rates of the black population, who were more likely to live in rural areas with no tuberculosis control programs.[38]

The history of the tuberculosis pandemic reveals much about the complexity and diversity of pandemics. The disease did not develop in persons shortly after their infection with the bacillus because a very large proportion of the population was infected in childhood but most cases of tuberculosis occurred among adults. The chronology of the pandemic and the patterns of mortality rates among population groups disprove the widely accepted explanation that the primary causal factors were deleterious social and hygienic conditions that resulted from urbanization and industrialization. The pandemic emerged in Massachusetts and England decades before these changes occurred and began to decline decades before conditions improved. Equally important, population groups that lived in similar urban neighborhoods and worked in similar industries varied substantially in their tuberculosis mortality rates. Although tuberculosis mortality rates were much lower in rural areas, the primarily rural black population in the United States had higher mortality rates at the peak of the pandemic than the more urban white population.

It is highly significant that public health experts in tuberculosis in the early twentieth century were disinclined to offer explanations for the decline of the pandemic. Researchers since that time have also found that the decrease in tuberculosis mortality rates defies a straightforward explanation. They all recognized its atypical, complex, and multifaceted nature.[39]

Conclusion

Pandemics are identified in terms of the relatively rapid rise and fall of their mortality rates in widespread geographic regions. This analysis found that many pandemics are also characterized by high mortality rates in population groups that are not typical victims of the normal disease. This was shown using the pandemics of influenza, lung cancer, and tuberculosis. Geographic variations in mortality rates also occur during pandemics. These findings indicate that pandemics can be understood only by analyzing trends in mortality rates by age, sex, and race groups and geographic locations throughout the pandemic. These types of analyses will be used to describe the emergence, peak period, and decline of the multinational pandemic of coronary heart disease in the mid twentieth century.

References

1. For descriptions of pandemics and endemics, see: J. N. Hays, *The Burdens of Disease: Epidemics and Human Response in Western History* Rev. Ed. (New Brunswick, NJ: Rutgers University Press, 1998); John Aberth, *Plagues in World History* (Plymouth, UK: Rowman & Littlefield, 2011); Alfred J. Bollet, *Plagues and Poxes: The Impact of Human History on Epidemic Disease* (New York: Demos, 2004).
2. William G. Rothstein, *Public Health and the Risk Factor: A History of an Uneven Medical Revolution* (Rochester, NY: University of Rochester Press, 2003), pp. 199-202, 240. For an overview of twentieth century health problems including many noninfectious diseases that can be considered as epidemics or pandemics, see Theodore M. Brown and Elizabeth Fee, "Social Movements in Health," *Annual Review of Public Health* 35 (2014): 385-98. The works cited contain references to additional research.
3. Godias J. Drolet, "Epidemiology of Tuberculosis," in *Clinical Tuberculosis*, ed. Benjamin Goldberg (Philadelphia, PA: Davis, 1947), pp. A5-A7; Gareth Williams, *Paralysed with Fear: The Story of Polio* (Houndsmills, Basingstoke, Hampshire, Eng.: Palgrave Macmillan, 2013); Allan M. Brandt, *Cigarette Century: The Rise, Fall, and Deadly Persistence of the Product that Defined America* (New York: Basic Books, 2007); Richard Kluger, *Ashes to Ashes: America's Hundred-Year Cigarette War, the Public Health, and the Unabashed Triumph of Phillip Morris* (New York: Vintage, 1997); Jonathan Engel, *The Epidemic: A Global History of AIDS* (New York: Smithsonian Books/Collins, 2006); D. S. Grimes, "An Epidemic of Coronary Heart Disease," *Quarterly Journal of Medicine* 105 (2012): 509-18.
4. For instances of historical epidemics throughout the world that were large enough to be noted, see George C. Kohn, *The Wordsworth Encyclopedia of Plague and Pestilence* (Ware, Hertfordshire, Eng.: Wordsworth, 1998). The experience in colonial America is described in John Duffy, *Epidemics in Colonial America* (Baton Rouge, LA: Louisiana State University Press, 1953).
5. Alfred W. Crosby, *America's Forgotten Pandemic: The Influenza of 1918* (New York: Cambridge University Press, 1989), p. 215; John M. Barry, *The Great Influenza: The Epic Story of the Deadliest Plague in History* (New York: Viking, 2004); Niall P.A.S. Johnson and Juergen Mueller, "Updating the Accounts: Global Mortality of the 1918-1920 'Spanish' Influenza Pandemic," *Bulletin of the History of Medicine* 76 (2002): 105-15.

6. George Dehner, *Influenza: A Century of Science and Public Health Response* (Pittsburgh, PA: University of Pittsburgh Press, 2012), pp. 53-55; Louis J. Dublin and Alfred J. Lotka, *Twenty-Five Years of Health Progress: A Study of the Mortality Experience among the Industrial Policyholders of the Metropolitan Life Insurance Company 1911 to 1935* (New York: Metropolitan Life Insurance Company, 1937), pp. 126-27. For a discussion of pathological issues, see Jeffery K. Taubenberger and David M. Morens, "1918 Influenza: the Mother of All Pandemics," *Emerging Infectious Diseases* 12 (2006): 15-22.
7. Crosby, *America's Forgotten Pandemic*, p. 203; Forrest E. Linder and Robert D. Grove, *Vital Statistics Rates in the United States, 1900-1940* (Washington, DC: Bureau of the Census, 1943), p. 125.
8. Locations in Europe and China have also been suggested as sources of the pandemic. See Barry, *The Great Influenza,* pp. 169, 453-56, and Aberth, *Plagues in World History,* pp. 114-15.
9. Dehner, *Influenza.* p. 43; Barry, *The Great Influenza,* p. 169; Crosby, *America's Forgotten Pandemic*, pp. 32, 62.
10. Dehner, *Influenza*, p. 47; Crosby, *America's Forgotten Pandemic*, p. 63.
11. Dehner, *Influenza*, p. 44; Crosby, *America's Forgotten Pandemic*, pp. 28-32, 56-63, 121-25.
12. Crosby, *America's Forgotten Pandemic*, p. 19; Dublin and Lotka, *Twenty-Five Years of Health Progress*, pp. 7-8.
13. Dublin and Lotka, *Twenty-Five Years of Health Progress,* p. 132.
14. Crosby, *America's Forgotten Pandemic*, pp. 215-16; Dehner, *Influenza*, p. 55.
15. Dublin and Lotka, *Twenty-Five Years of Health Progress,* p. 132.
16. Crosby, *America's Forgotten Pandemic*, p. 132.
17. Dublin and Lotka, *Twenty-Five Years of Health Progress,* p. 129.
18. Dublin and Lotka, *Twenty-Five Years of Health Progress,* p. 129.
19. Barry, *The Great Influenza* p. 169-70; Dehner, *Influenza*, p. 44.
20. Dublin and Lotka, *Twenty-Five Years of Health Progress,* p. 158.
21. Richard Doll, "Occupational Lung Cancer: A Review," *British Journal of Industrial Medicine* 16 (1959): 181-90.
22. For an analysis of cigarette smoking as an epidemic, see Alan D. Lopez, Neil E. Collishaw, and Tapani Piha, "A Descriptive Model of the Cigarette Epidemic in Developed Countries," *Tobacco Control* 3 (1994): 242-47.
23. Rothstein, *Public Health and the Risk Factor*, p. 240-41.
24. Robert D. Grove and Alice M. Hetzel, *Vital Statistics Rates in the United States, 1940-1960* (Washington, DC: National Center for Health Statistics, 1968), pp. 408-12.
25. Brandt, *The Cigarette Century,* pp. 27-31, 53, 86-87; *The Tax Burden on Tobacco: Historical Compilation* 46 (2011): 6.
26. National Center for Health Statistics, *Health, United States, 2013* (Hyattsville, MD: 2104), pp. 117-18
27. Rothstein, *Public Health and the Risk Factor*, pp. 238-59.
28. Drolet, "Epidemiology of Tuberculosis," p. A3.
29. Drolet, "Epidemiology of Tuberculosis," p. A3.

30. John Charlton, Patricia Fraser, and Mike Murphy, "Medical Advances and Iatrogenesis," in *The Health of Adult Britain, 1841-1994,* ed. John Charlton and Mike Murphy, 2 vols. (London: Office for National Statistics, 1997), I:221; U.S. Bureau of the Census, *Historical Statistics of the United States. Colonial Times to 1970,* 2 vols. (Washington, DC: 1975), p. I:63; Drolet, "Epidemiology of Tuberculosis," pp. A5-A7.
31. Charlton, Fraser, and Murphy, "Medical Advances and Iatrogenesis," p. I:221; U.S. Bureau of the Census, *Historical Statistics of the United States,* p. I:63.
32. John Charlton, "Trends in All-Cause Mortality," in *The Health of Adult Britain,* p. I:23; U.S. Bureau of the Census, *Historical Statistics of the United States,* p. I:57.
33. Drolet, "Epidemiology of Tuberculosis," p. A-15.
34. Rothstein, *Public Health and the Risk Factor,* pp. 95, 103-18; Karl Pearson, *The Fight against Tuberculosis and the Death-rate from Phthisis,* (Cambridge, Eng.: Cambridge University Press, 1911), pp. 20-27; Drolet, "Epidemiology of Tuberculosis," pp. A23-27, A53.
35. W. H. Frost, "How Much Control of Tuberculosis?" *American Journal of Public Health* 27 (1937): 759-60; Drolet, "Epidemiology of Tuberculosis," p. A18.
36. Drolet, "Epidemiology of Tuberculosis," pp. A21-23, A5-7.
37. U.S. Bureau of the Census, *Historical Statistics of the United States,* p. I:12.
38. Frost, "How Much Control of Tuberculosis?" pp. 563-64. For evidence of the ineffectiveness of programs to control tuberculosis in the early twentieth century, see Rothstein, *Public Health and the Risk Factor,* pp. 108-18.
39. Gerald N. Grob, *The Deadly Truth: A History of Disease in America* (Cambridge, MA: Harvard University Press, 2002), pp. 211-215; Hayes, *The Burdens of Disease,* p. 161.

Chapter 2

Overview of the Coronary Heart Disease Pandemic in the United States and Methods of Analysis

This chapter describes broad aspects of the coronary heart disease pandemic that will be examined in detail subsequently. The emergence of coronary heart disease as a major cause of morbidity and mortality in the 1930s aroused great concern among health professionals. Experts considered it to be a known disease that had become more severe and spread throughout the population because of changes in diets and other aspects of lifestyles in advanced societies. High disease rates would continue indefinitely unless the diets and lifestyle were modified. The increased number of cases of the disease led to greater understanding of its physical characteristics and changes in the disease classification system. The experts developed theories to explain the causes of the disease using a probabilistic concept of disease etiology known as the risk factor. The methods used in this study to analyze coronary heart disease mortality rates during the pandemic are described.

Overview of the Coronary Heart Disease Pandemic

Coronary heart disease, which has undergone several changes in terminology described below, produced more adult deaths in most advanced countries than any other disease during much of the twentieth century. Early in the twentieth century it was a minor cause of death, primarily among the elderly, that was believed to be caused by "hardening of the arteries" as people aged. About 1930 a different type of coronary heart disease emerged that affected all adult age groups in most of the wealthiest countries in the world. Mortality rates reached their peak from the 1950s to the 1970s and then declined steadily during the last decades of the century. The disease became a major cause of death during a period of improvements in the standards of living of the countries and decreases in their adult mortality rates from all other causes.[1]

Coronary heart disease results from a total or partial blockage of blood flow through one or more of the coronary arteries that provide blood to the heart muscle. The loss of blood flow to the muscle can stop or reduce the heart's pumping action and produce death quickly or after some period of time. Those who survive an episode of the disease are at greater risk of future serious heart problems, including additional blockages and heart failure. An abrupt and severe blockage is technically called a myocardial infarction and colloquially a heart attack in the United States.

The severity of the coronary heart disease pandemic using United States vital statistics is indicated by estimates of the proportion of all adult deaths that were caused by the disease in its early period, its peak period, and after it had been declining for several decades (Table 8.1). Mortality rates varied considerably by age, sex, and race, so that it is necessary to examine each population group separately. The disease was classified as angina pectoris and diseases of the coronary arteries in 1940 and as ischemic heart disease in 1970 and subsequently. Considering those ages 55-64, among white men coronary heart disease mortality constituted 15 percent of all deaths in 1940, 41 percent in 1970, 26 percent in 1990, and 18 percent in 2010. Among white women the proportions of all deaths caused by coronary heart disease increased from 7 percent in 1940 to 27 percent in 1970 and then decreased to 16 percent in 1990 and 10 percent in 2010. Among black men the proportions were 5 percent in 1940, 29 percent in 1970, and 16 percent in both 1990 and 2010. Among black women, the proportions were 3 percent in 1940, 30 percent in 1970, 16 percent in 1990, and 12 percent in 2010.[2]

The coronary heart disease pandemic emerged during the 1930s when infectious diseases, the major causes of death throughout history, continued to be the focus of attention of both health professionals and the public in advanced countries. Mortality rates from infectious diseases had decreased substantially in the early twentieth century, and this was recognized as a great achievement and closely monitored. The sulfa drugs in the 1930s and penicillin in the 1940s revolutionized the treatment of bacterial infectious diseases and aroused great interest in future innovations that could improve the treatment of all infectious diseases.[3]

A focus on infectious diseases also continued because of frightening outbreaks of certain infectious diseases in the early twentieth century. The great influenza pandemic from 1917 to 1920 produced tens of millions of deaths in both rich and poor countries throughout in the world as described in Chapter 1. The much smaller pandemic of paralytic polio that produced deaths and disabilities in young children began somewhat later in the United States and other advanced countries. The fundamental characteristics of both diseases were poorly understood, which only confirmed the need for continuing vigilance and intensified research on infectious diseases.[4]

As a result, the emergence of the coronary heart disease pandemic did not receive the attention that it merited in the United States and other advanced countries. The decrease in mortality rates from infectious diseases enabled more people to live to old age and it seemed inevitable that their hearts should fail late in their lives. It was a source of legitimate pride for physicians and public health officials to know that people died at older ages instead of infancy, childhood, or early adulthood. Physicians in the new specialty of cardiology paid primary attention to other heart problems that occurred to the increasing number of elderly persons and were more amenable to treatment. In his history of cardiology, Fye observed that the specialists developed a particular interest in coronary heart disease in the 1950s and 1960s, when prominent politicians and others experienced heart attacks. The interest was strengthened by the enactment in 1965 of Medicare, which provided federal funding for medical care of the elderly.[5]

The coronary heart disease pandemic also received little attention in the United States because most health professionals and the public did not believe that it was a pandemic. Rather, high coronary heart disease rates were believed to be the inevitable and unavoidable result of persons living to older ages because of better public health and medical care that reduced mortality rates from infectious diseases. These older persons were more susceptible to coronary heart disease than in the past because modern lifestyles weakened human vitality and vigor.[6]

The lifestyle theory of the causes of the rapid increases in coronary heart disease rates dominated medical thinking in the second half of the twentieth century. The components of lifestyles that were of greatest concern included diets, particularly fat intake, overweight, physical inactivity, and cigarette smoking. Many experts believed that the high coronary heart disease rates could be reduced only by modifying these lifestyle factors. They considered modern diets to be particularly important, and this became known as the "diet-heart" hypothesis. The experts claimed as essential for good health foods that people in advanced countries had gladly stopped eating as their incomes rose. They pointed with admiration to southern European countries with low coronary heart disease mortality rates and impoverished and backward economies.[7]

The experts' fatalistic acceptance of high rates of coronary heart disease as the result of lifestyles in modern societies produced amazement and confusion among them when coronary heart disease mortality rates began to decrease steadily and substantially in the 1970s and 1980s in the United States and other advanced countries. Two participants at a 1978 medical conference on the decline in coronary heart disease sponsored by the National Institutes of Health observed: "The announcement for this reversal in the long-term upward trend was received with great astonishment, both in the United States and other countries."[8]

Despite the decreases in coronary heart disease mortality rates, many experts at the turn of the twenty-first century continued to insist that modern lifestyles have been and continue to be the major causes of coronary heart disease.[9] They have promoted the diets of four-legged herbivores and other lifestyle modifications as proven methods of preventing coronary heart disease. They have strongly advocated the widespread use of drugs that lowered risk factors for coronary heart disease even though the much lower disease rates have made the drugs beneficial for a much smaller proportion of the population.

This study will provide an alternative history of coronary heart disease in the last two-thirds of the twentieth century. It will demonstrate that the increases and decreases of coronary heart disease mortality rates constituted a multinational pandemic comparable to other recent great pandemics described in Chapter 1. As with other great pandemics, the characteristics of the patients who experienced pandemic coronary heart disease differed substantially from those who experienced the normal disease before and after the pandemic. This study will examine trends and patterns of coronary heart disease mortality rates of age, sex, and race groups in the United States and age and sex groups in other countries that experienced the pandemic. It will examine geographic variations in coronary heart disease mortality rates within and among countries during the pandemic.

This study will contend that is inconceivable that the many advanced countries on three continents that experienced the pandemic underwent identical changes in their diets and lifestyles at the same times before the emergence of the pandemic and identical reverse changes in their diets and lifestyles at the same times before the decline of the pandemic. It will examine changes in individual lifestyle risk factors in the United States during the emergence and decline of the pandemic. It will show a lack of correspondence between the trends in these alleged causal factors and the trends in coronary heart disease mortality rates during the emergence and decline of the pandemic in the United States.

This study will demonstrate that the coronary heart disease pandemic was caused by factors that emerged at a particular time and diffused rapidly among many advanced countries in the world. It will also demonstrate that coronary heart disease mortality rates began to decrease at about the same times in all of the countries. The factors were novel because of the many differences between the victims of the pandemic disease and the normal disease. This analysis is descriptive and does not identify the responsible factors. It is worth observing that it took a half century for an explanation of the causes of the great influenza pandemic and several decades for explanations of the causes of the polio and lung cancer pandemics.

Characteristics of and Risk Factors for Coronary Heart Disease

Coronary heart disease is characterized by a partial or total blockage of the flow of blood in one or more of the coronary arteries that supply blood to the heart muscle or myocardium. The lack of a blood supply to the heart or any body part is termed ischemia, which led eventually to the term ischemic heart disease. A partial blockage of blood flow to the heart muscle that produces chest pain, often after physical exertion or emotional stress, is called angina pectoris. A partial or total blockage that produces the necrosis or death of some tissue (infarct) in the heart muscle (myocardium) is called a myocardial infarction, or heart attack in popular terms. Acute myocardial infarctions can produce death quickly or after longer periods of time, often from subsequent attacks of the disease or other diseases to which the person has become more susceptible. The necrosis cannot be healed but the development of scar tissue can permit the patient to return to varying degrees of normal life.[10]

The causes of the blockages have been explained by two different theories that have been the subject of considerable controversy. One theory holds that blockages are typically caused by a blood clot (thrombus) that develops in a coronary artery and blocks the flow of blood to the heart muscle. The other theory holds that atherosclerosis, a mass termed an atheroma, forms on the inner wall of a coronary artery. The mass fills the hollow interior (lumen) of the artery partially or completely, which restricts or blocks blood flow through the artery. Atheromas contain fibrin, blood platelets, cholesterol, collagen, and other substances. Clots can result from the breakdown of atheromas, although the frequency with which this occurs is highly controversial.[11]

Coronary heart disease in the form of angina pectoris was identified and labeled in the late eighteenth century and was accurately described in the medical literature in the nineteenth and early twentieth centuries. Although it was uncommon in the nineteenth century, its distinctive symptoms made it easily diagnosed. An accurate description of myocardial infarction did not occur until the early twentieth century after a period of research to understand the condition and its relationship to angina pectoris. Its general acceptance in the 1930s was facilitated by the use of the electrocardiogram.[12]

Because no specific causal factor is necessary or sufficient for the development of coronary heart disease (or any chronic disease), the etiology of the disease must be conceived as multifactorial. The objective has been to identify characteristics of individuals that can increase their probability of the future occurrence of the disease. Each characteristic is known as a risk factor, a concept and term developed by the life insurance industry early in the twentieth century to predict the probability of premature mortality of applicants for policies. The concept and term were adopted by the Framingham Heart Study, which began in 1948 and was the most important and methodologically sophisticated early heart disease research project. According to George V. Mann, one of the early Framingham researchers, "In the 1950s, . . . Roy Dawber, Felix Moore, and I began to use the phrase 'risk factor,' a term adopted from the insurance industry." The researchers selected more than 5000 residents of Framingham, Massachusetts using scientific sampling methods and studied risk factors for coronary heart disease in their sample by following them for many years. The great influence of the Framingham study and the usefulness of the risk factor concept has led to its adoption for many diseases.[13]

Risk factors are individual behaviors or physiological conditions that increase the future probability of disease. Individual risk factors vary greatly in their occurrence in the population. They also vary in the probability of future occurrence of disease and in the length of time necessary for disease to develop. Widely accepted risk factors for coronary heart disease that involve individual behaviors are typically strongly influenced by social conditions, including diet, physical inactivity, and cigarette smoking. Other risk factors that involve physiological conditions, including high blood pressure, high blood cholesterol, and obesity, also have social components.[14]

In general, the more extreme the level of a person's risk factor relative to the normal level, the greater the risk of future disease. However, only a very small proportion of persons with risk factors have extreme levels of them. The great majority of persons with risk factors have levels that differ from normal levels by small to moderate amounts. As a result, most persons with risk factors have low probabilities of the development of future disease.

The levels of each risk factor deemed extreme enough to warrant intervention are always matters of judgment, never scientific conclusions. Some individuals or groups may consider a particular level of a risk factor to be a serious health problem while others may consider it to be of lesser importance. Government agencies, professional societies, and voluntary health organizations within and among countries differ greatly among themselves in these judgments.

Coronary Heart Disease and Disease Classification Systems

Improved knowledge of coronary heart disease has led to changes in its terminology and classification. The standard disease classification system was adopted in the 1890s and called the International List of Causes of Death. It has been revised periodically and the title changed to the International Classification of Diseases. Responsibility for the revisions were assumed after the mid twentieth century by the World Health Organization. The revision dates used in this study are those of the organizations that created and approved the revisions. Individual countries have varied in the dates of their adoption of each revision.[15]

A key issue in the early history of the development of disease categories was whether to enumerate diseases by the pathological process or the anatomical site involved. For example, knowledge of the tubercle bacillus led to the decision to group all cases of tuberculosis into a single category and subdivide them by site, rather than list tuberculosis of each site as a subdivision of the diseases of that site. The need for this type of decision had important implications for coronary heart disease. The several pathological processes involved in coronary heart disease were originally listed in separate disease categories that were not limited to a particular organ. As coronary heart disease became more important as a cause of death, decisions were made over several revisions to transfer the subcategories of the relevant separate disease categories into a single category based on the anatomical site.[16]

This process can be seen by examining the revisions of the disease classification system for coronary heart disease. "Angina pectoris," a common symptom of coronary heart disease, was the term used in the first revision (1900-09), second revision (1910-20) and third revision (1921-1929). The fourth revision (1930-1938) and the fifth revision (1939-1948) combined "angina pectoris" with another category, "diseases of the coronary arteries," which was a pathological process. The latter category was taken from subcategories of two other disease categories, "diseases of the arteries" and "embolism and thrombosis," which were not limited to the heart. The sixth revision (1949-1957) and seventh revision (1958-1967) introduced a new term, "arteriosclerotic heart disease, including coronary disease." This new term combined two terms that had been listed in different categories in the fifth revision: "angina pectoris and disease of the coronary arteries" and "arteriosclerotic heart disease," which had been listed under "chronic myocarditis and myocardial degeneration." The eighth revision (1968-78) introduced the new term "ischemic heart disease," which was primarily but not totally a continuation of the seventh revision. The ninth edition (1979-94) and tenth edition (1995-) continued to use ischemic heart disease with minor modifications.[17]

These changes were part of a steady expansion of the number of categories devoted to all forms of heart diseases as they became more important causes of death and better understood because of more older persons in the population. Only four categories were used for heart diseases in the first three revisions. These increased to 13 in the 1930 revision, 20 in the 1939 revision, and 41 in the 1955 revision.[18] The periodic revisions of the coronary heart disease category demonstrate that it was becoming a greater source of concern to physicians and public health officials.

This study will analyze trends in coronary heart disease mortality rates within the time period covered by a single revision whenever possible to avoid problems caused by changes in terminology. When a new category combines deaths from more than one previous category, it increases the number of enumerated deaths regardless of any actual increase in the number of deaths. When patients die of multiple causes, they are assigned an underlying cause of death. The number of deaths can change when revisions change the underlying cause or the method used to assign the underlying cause.

Another consequence of changes in terminology is that physicians adopt the new terms gradually. A historical study of death certificates in a city near Boston, Massachusetts in 1936 found an increase in deaths from "angina pectoris" from 1900 to 1930 and then a decrease to 1935 while deaths increased rapidly from the new 1930 category, "diseases of the coronary arteries." The large increase in the number of deaths in the new category resulted from physicians preferring it to angina pectoris plus the many new cases in the growing pandemic.[19]

The changes in disease categories did not mean that angina pectoris had been problematic as a disease concept. A British physician stated in 1946 that the recent increase in coronary heart disease rates was real because its clinical symptoms were described in the late eighteenth century and "angina pectoris is one of the easiest of all diseases to recognize." He examined the case records of several eminent physicians at the turn of the twentieth century who "were at least as competent to diagnose angina pectoris as are physicians of this generation" and found that they reported very few cases of angina pectoris. He said that thrombosis or blockage of a coronary artery was a greater diagnostic problem because the "clinical picture" was much less clear. He attributed the rapid adoption of coronary thrombosis as a diagnostic term after 1930 in part to greater recognition of its clinical symptoms by American physicians in the 1920s.[20]

The ability of physicians to recognize coronary heart disease during the early twentieth century was shown in reviews in the mid-1920s of articles in medical journals. The reviews found that the authors had an accurate knowledge of the clinical manifestations of coronary heart disease and its epidemiological features. A common manifestation of coronary heart disease is sudden death, and this was sufficiently distinctive to be readily diagnosed.[21]

Methods of Analysis

Pandemics of chronic diseases in the twentieth century have usually been difficult to measure and analyze. They must be differentiated from normal disease patterns as well as from diseases that have increased in prevalence with the increase in the proportion of older persons in the population.

The emergence and decline of pandemics have traditionally been measured by trends in mortality rates in government vital statistics because the diagnoses listed on death certificates are more accurate and available than reports of diseases in living patients. The early twentieth century witnessed a revolution in the use of death certificates based on diagnostic categories developed by an international organization and adopted by all advanced countries. Advanced countries required death certificates for all deaths to be filed with government authorities, which permitted the accurate enumeration of mortality statistics for the first time in history. Comprehensive population censuses produced accurate enumerations of populations and population groups so that it was possible to calculate mortality rates for the total population and age, sex, race, and other groups. The coronary heart disease pandemic was one of the first major pandemics to benefit from this quantification of government vital statistics.[22]

In analyzing mortality trends from a disease over time, a key issue is whether the proportion of misdiagnosed cases is increasing or decreasing. So long as the proportion remains relatively constant, the trends over time will usually indicate actual changes in mortality rates. It will be shown that the coronary heart disease pandemic was characterized by much higher mortality rates of white men than white women in the middle age groups and moderately higher rates in the older age groups. It will also be shown that this difference did not occur for other diseases of the heart or circulatory system. Thus a useful measure of the accuracy of coronary heart disease mortality statistics during the pandemic is a comparison of white male death rates relative to white female death rates in different age groups. Enumerations of deaths in relevant disease categories during the pandemic that do not show this distinctive pattern cannot have been caused by coronary heart disease.

Because age is the most important factor affecting mortality, it is inappropriate to compare mortality rates in total populations that differ in their age distributions. Coronary heart disease mortality rates for men were higher than those for women, so that it is also necessary to examine the mortality rates of each sex separately. The method used whenever possible in this study is to compare mortality rates for the same age and sex groups at different time periods or geographic locations. A widely used but less accurate alternative method is to make the age distributions of different populations conform to an arbitrary standard age distribution using a method called age-adjustment or age-standardization All age-adjusted comparisons of mortality rates among states in the United States in this study use the 1940 standard United States population. Comparisons of the age adjusted mortality rates of groups are appropriate only when they are based on the same standard population. This often limits the ability to make comparisons among countries and time periods.

The mortality trends of a pandemic disease must be differentiated from mortality trends from all other causes to demonstrate that the pandemic trends were not part of general changes in health and illness. Total mortality rates excluding coronary heart disease will be examined for all population groups being studied throughout the pandemic and compared to their coronary heart disease mortality rates. Trends in mortality rates for coronary heart disease will also be compared to trends for other diseases involving the circulatory system.

This study will analyze the coronary heart disease pandemic as a multinational phenomenon by first examining its development in the United States and then in other advanced countries. The United States has many advantages as a source of data. The quantity, quality, and availability of its historical vital statistics during all except the early years of the twentieth century are, in the opinion of the author, superior to those of any other country. Its populations of 132 million persons in 1940, 151 million in 1950, and 282 million in 2000 were as large as the combined population of many other countries. The large populations permit more accurate measurement of the patterns of the pandemic and mortality rates of age, sex, race, and other population groups. Medical education in the United States was reasonably standardized by the 1930s so that all medical students learned the same methods of diagnosis and reporting.[23]

Geographic variations in coronary heart disease mortality rates were an important characteristic of the pandemic, which is an additional advantage of studying the United States. Age-adjusted coronary heart disease mortality rates for men and women are compared beginning in 1950 for the 48 states in the United States at that time. Comparisons among the states have many methodological advantages over comparisons among countries. The states have the same methods of diagnosing disease and measuring mortality and share many similarities in their standards of living, health care systems, and lifestyles. The states varied considerably in their coronary heart disease mortality rates and in many measured characteristics that have a direct impact on health and illness. This permits an analysis of the degree of correspondence between the coronary heart disease mortality rates of the states and other quantifiable characteristics of the states using Pearson correlation coefficients. The extent of the variations among the states for each characteristic will be measured using standard deviations.

An issue that arises when comparing mortality rates among population groups is that the comparisons utilize rates, averages, or other characteristics of population groups as the unit of analysis, which are called "ecological correlations." Individuals, not states or countries, develop coronary heart disease so that it is necessary to justify comparisons of coronary heart disease mortality rates among states or countries rather than individuals. Pandemics develop in and diffuse among geographic regions because of social factors, many of which operate at the levels of states or countries. For this reason it is scientifically valid to analyze the effects of these types of social factors by comparing the coronary heart disease mortality rates of states or countries.

The descriptions of pandemics in Chapter 1 indicate the need for two types of initial analyses of the coronary heart disease pandemic. One is the amount of the general increases in coronary heart disease mortality rates over time and the dates of their occurrence. A second is differences in the increases in mortality rates for individual age, sex, and race groups. If overall mortality rates increased much more than in normal times and the greatest increases occurred among groups that did not have high rates of the normal disease, this indicates that a new type of coronary heart disease was responsible for the changes.

References

1. For perspectives of coronary heart disease in the twentieth century similar to this study, see Gerald N. Grob, "Coronary Heart Disease and Cancer in Twentieth-Century America: An Etiological Dilemma," *New Jersey Medicine* 99 (2002):29-37 and D.S. Grimes, "An Epidemic of Coronary Heart Disease," *Quarterly Journal of Medicine* 105 (2012):509-18.
2. See Table 4.1 and Table 8.1.
3. For a description of infectious diseases in the United States in the first half of the twentieth century, see Gerald N. Grob, *The Deadly Truth: A History of Disease in America* (Cambridge, MA: Harvard University Press, 2002), pp. 180-216 and Harry F. Dowling, *Fighting Infection: Conquests of the Twentieth Century* (Cambridge, MA: Harvard University Press, 1977).
4. George Dehner, *Influenza: A Century of Science and Public Health Response* (Pittsburgh, PA: University of Pittsburgh Press, 2012); Gareth Williams, *Paralysed with Fear: The Story of Polio* (Hampshire, England: Palgrave Macmillan, 2013).
5. W. Bruce Fye, *American Cardiology: The History of a Speciality and its College* (Baltimore: Johns Hopkins University Press, 1996), pp. 174, 188-89, 216, 223-227.
6. William G. Rothstein, *Public Health and the Risk Factor: A History of an Uneven Medical Revolution* (Rochester, NY: University of Rochester Press, 2003), pp. 208-9; Daniel Levy and Susan Brink, *A Change of Heart: How the Framingham Heart Study Helped Unravel the Mysteries of Cardiovascular Disease* (New York: Knopf, 2005), pp. 22-34.
7. National Center for Health Statistics, *Healthy People 2000 Review, 1998-99* (Hyattsville, MD: Public Health Service, 1999), pp. 148-49; Harvey Levenstein, *Fear of Food: A History of Why We Worry about What We Eat* (Chcago, IL: University of Chicago Press, 2012), pp. 125-159.
8. Harry M. Rosenberg and A. Joan Klebba, "Trends in Cardiovascular Mortality with a Focus on Ischemic Heart Disease," in *Proceedings of the Conference on the Decline in Coronary Heart Disease Mortality*, ed. Richard J. Havlik and Manning Feinleib (Bethesda, MD: National Institutes of Health, 1978), p. 15.

9. For a description of the status of coronary heart disease in the early twenty-first century, see Fabian Sanchis-Gomar, et al, "Epidemiology of Coronary Heart Disease and Acute Coronary Syndrome," *Annals of Translational Medicine* 4 (2016): 256-67.
10. Robert A. Aronowitz, *Making Sense of Illness: Science, Society, and Disease* (Cambridge, Eng.: Cambridge University Press, 1998). pp. 84-110; Melvin L. Marcus, *The Coronary Circulation in Health and Disease* (New York: McGraw-Hill, 1983), pp. 65-66, 261, and passim.
11. Rothstein, *Public Health and the Risk Factor*, pp. 286-90; Allen B. Weisse, "The Elusive Clot: The Controversy over Coronary Thrombosis in Myocardial Infarction," *Journal of the History of Medicine and Allied Sciences* 61 (2006): 66-78; N.G.B. McLetchie, "The Pathogenesis of Atheroma," *American Journal of Pathology* 28 (1952): 413-35; M. Daria Haust, Robert H. More, and Henry Z. Movat, "The Mechanism of Fibrosis in Atherosclerosis," *American Journal of Pathology* 35 (1959): 265-73. An examination in 2015 of the web sites of major American public and private organizations concerned with health policy and health care shows substantial disagreements about the causes of coronary occlusion. In addition, these web sites place a much greater emphasis on the cholesterol content of atheromas than did the experimental pathologists who conducted the fundamental research in the mid twentieth century. For an example of the early research, see J.B. Duguid, "The Etiology of Atherosclerosis," *Practitioner* 175 (1955): 241-47.
12. William J. Proudfit, "Origin of Concept of Ischemic Heart Disease," *British Heart Journal* 50 (1983): 209-12; W. Bruce Fye, "Acute Myocardial Infarction: A Historical Summary," in *Acute Myocardial Infarction*, ed. Bernard J. Gersh and Shahbudin H. Rahimtoola 2nd ed. (New York: Chapman and Hall, 1997).
13. Rothstein, *Public Health and the Risk Factor*, pp. 61-66, 279-85; Levy and Brink, *A Change of Heart*; George V. Mann, "The Clinical Trials," in *Coronary Heart Disease: The Dietary Sense and Nonsense*, ed. George V. Mann (London Eng: Janus, 1993), p. 74.
14. Rothstein, *Public Health and the Risk Factor*, pp. 2-5.
15. Iwao M. Moriyama, Ruth M. Loy, Alastair H.T. Robb-Smith, *History of the Statistical Classification of Diseases and Causes of Death*, ed. Harry M. Rosenberg and Donna L. Hoyert (Hyattsville, MD: National Center for Health Statistics, 2011), pp. 9-22.
16. Moriyama, Loy, and Robb-Smith, *History of the Statistical Classification of Diseases and Causes of Death*, p. 15.
17. Moriyama, Loy, and Robb-Smith, *History of the Statistical Classification of Diseases and Causes of Death*, pp. 9-21; Iwao M. Moriyama, Dean E. Krueger, and Jeremiah Stamler, *Cardiovascular Diseases in the United States* (Cambridge, MA: Harvard University Press, 1971), pp. 31-37; John Charlton and Mike Murphy, eds., *The Health of Adult Britain, 1841-1994*, 2 vols. (London: Office for National Statistics, 1997), II:211-12.
18. Moriyama, Krueger, and Stamler, *Cardiovascular Diseases in the United States*, p. 32.
19. Francis Denny, "The Increase in Coronary Disease and its Cause," *New England Journal of Medicine* 214 (1936): 769-73.

20. Maurice Cassidy, "Coronary Disease: The Harveian Oration of 1946," *Lancet* Oct. 26, 1946: 587-90. For a discussion of the recognition of coronary heart disease in the early years of the pandemic, see Rothstein, *Public Health and the Risk Factor*, pp. 195-209.
21. J.W. McNee," The Clinical Syndrome of Thrombosis of the Coronary Arteries," *Quarterly Journal of Medicine* 19 (1925):44-49; G.A. Allan, "Diseases of the Coronary Arteries," *British Medical Journal* 2 (1928):232-38.
22. Moriyama, Loy, and Robb-Smith, *History of the Statistical Classification of Diseases and Causes of Death.*
23. William G. Rothstein, *American Medical Schools and the Practice of Medicine: A History* (New York: Oxford University Press, 1987), pp. 153-78.

Chapter 3

The Emergence of the Coronary Heart Disease Pandemic in the United States, 1910-1935

Coronary heart disease changed from an unimportant cause of death in the early twentieth century to become a major cause of death in the 1930s. The new pandemic coronary heart disease was strikingly different from the normal disease. It produced much higher mortality rates in all adult age, sex, and race groups. It produced greater increases in the mortality rates of men than women and older than younger age groups in both the white and black populations. The increases in coronary heart disease mortality rates occurred during a period when no increases occurred in mortality rates from other important chronic diseases that affected the heart and arteries. These changes are described using an analysis of the mortality rates of millions of policyholders of the Metropolitan Life Insurance Company.

The analyses of the pandemics of influenza, lung cancer, and tuberculosis in Chapter 1 demonstrated that the population groups that developed the pandemic form of the disease differed significantly from those that developed the disease in normal times. The influenza pandemic produced striking increases in mortality rates among young adults even though influenza was normally a disease of the very young and the very old. The lung cancer pandemic initially increased lung cancer mortality rates among men who were never exposed to the known carcinogens for the disease. Tuberculosis mortality rates during the peak of the pandemic varied among population groups in different ways than after the pandemic.

Similar developments occurred with the emergence of the coronary heart disease pandemic in the United States. Early in the century coronary heart disease was an unimportant cause of death among elderly men and women and was believed to be one of several consequences of a condition known as "hardening of the arteries." During the 1920s a new kind of coronary heart disease increased mortality rates among younger as well as older adults and produced much greater increases in the mortality rates of men than women. Physicians considered it extremely improbable that hardening of the arteries could have caused these increases given the young ages of some victims and the improved general health of the population, including the elderly.[1]

Research in the second quarter of the twentieth century demonstrated that diagnoses of coronary heart disease were generally accurate and that the increase could not be explained by misdiagnoses. Experts in diseases of the heart agreed that coronary heart disease had been very rare early in the century and that the increases were so large that misdiagnoses would have produced decreases in mortality rates from other diseases of the heart and arteries, which did not occur. Some studies used physiological measures, such as a review of records of 2877 autopsies at a hospital in New York City published in 1934. It found 762 autopsies that mentioned diseases of the coronary arteries from 1910 to 1931 and that the number with coronary heart disease increased steadily over the period. Anderson and LaRiche also found that the increase in coronary heart disease mortality was much greater than for other diseases with which it could be confused. They reclassified samples of between 2500 to 5000 death certificates of men ages 45-64 in Ontario in each Canadian census year from 1901 to 1961. They used three definitions of heart disease: a narrow definition limited to coronary heart disease, a broader definition that included other forms of heart disease, and a very broad definition that added other causes of death that might be confused with coronary heart disease. They found that much greatest increases in mortality rates occurred for the narrow definition. These findings clearly showed that the increase in coronary heart disease mortality rates was real.[2]

One of several vital statistics measures that demonstrated that this was a new form of coronary heart disease was the much higher mortality rates among men than women. The sex difference was much smaller early in the century, so that it was not a characteristic of normal coronary heart disease. No sex difference occurred before or during the pandemic for other diseases of the heart and arteries, as will be demonstrated below. If the very large increase in coronary heart disease mortality rates among men had been due to misdiagnoses, mortality rates of men from other diseases of the heart and arteries would have decreased relative to those of women. This did not occur.

Increases in Coronary Heart Disease Mortality Rates

Enumerations of the increases in mortality rates from coronary heart disease for all age groups during the early twentieth century pose several problems. They include the revisions in disease categories described in Chapter 2 as well as the lack of availability of federal government vital statistics for the total population and individual population groups. In 1900 the United States federal government established the Death Registration Area to create standardized national mortality data. This was necessary because states varied in their methods of gathering mortality data, and many states gathered them haphazardly or not at all. The Death Registration Area accepted data only from states considered to have accurate reporting. The number of reporting states increased from 25 in 1915 to 35 in 1920, 47 in 1929, and all 48 states in 1933.[3]

The small number of states in the Death Registration Area early in the century and questions about their representativeness of the total population indicate the need to utilize other sources of national mortality statistics. Probably the most useful national data to measure coronary heart disease mortality trends from 1911 to 1935 are the statistical analyses of the deaths of policyholders ages 1-74 who owned industrial life insurance policies sold by the Metropolitan Life Insurance Company. Practically all deaths and causes of deaths were reported because a death certificate was required for payment of the death benefit. Industrial life insurance was an extremely popular form of life insurance designed to cover burial expenses for low income persons living in densely populated urban neighborhoods. Premiums of as little as a few cents weekly were collected by agents who walked the streets and visited the homes of the policyholders.[4]

The millions of Metropolitan industrial policyholders provide a useful measure of national changes in urban coronary heart disease mortality rates in the United States over this period. The number of policyholders increased from 8 million in 1911 in a United States population of 94 million to 17 million in 1935 in a population of 127 million. Thus the policyholders constituted 9 percent of the United States population in 1911 and 13 percent in 1935. In 1934 the policyholders were almost totally urban, with only 5 percent of men employed in agriculture, forestry, or animal husbandry, and in lower socioeconomic groups, with 52 percent employed in manufacturing, 13 percent in transportation, and only 2 percent in professional or semiprofessional occupations. White women accounted for 50 percent of policyholders, white men 39 percent, colored (the term used by the researchers) men 5 percent, and colored women 6 percent. Considering policyholders old enough to be victims of coronary heart disease, in 1934 34 percent of the policyholders were ages 35-74, compared to 38 percent of the United States population.[5]

A study of the mortality rates of Metropolitan Life Insurance Company industrial policyholders has major advantages. Most policyholders were low income residents of urban areas, which were among the first population groups affected by the emergence of the pandemic, as will be shown in Chapter 4. The millions of policyholders permit the computation of accurate mortality rates for race and sex groups within each age group. The similar backgrounds and the good health of the policyholders when they purchased the policies greatly reduce the possibility that changes in the characteristics of the policyholders affected the trends in mortality rates.

The analyses were performed by Louis I. Dublin and Alfred J. Lotka, two Metropolitan Life Insurance Company employees who were among the most eminent American statisticians who studied mortality. Both men were authors of scholarly articles and books and each served as president of the American Statistical Association and the Population Association of America. Dublin also served as president of the American Public Health Association and Lotka is considered one of the founders of mathematical demography. They published many statistical analyses of policyholder health and disease that were used by public health departments, government agencies, medical researchers, newspapers, and others as the best available information on national patterns of health and disease. Their methods of analysis of policyholder mortality were at the forefront of statistical research on mortality.[6]

The research described here examined 3.2 million deaths of Metropolitan Life Insurance Company industrial life insurance policyholders ages 1 to 74 from 1911 to 1935. The average annual death rate standardized for age, sex, and race was 8.9 deaths per 1000 policyholders for the period from 1920 to 1934, which was higher than the United States crude death rate of 7.7 per 1000 persons. The higher Metropolitan overall death rate was characteristic of lower income urban Americans generally.[7]

The period from 1911-15 to 1931-35 saw substantial improvements in the health of the policyholders (Table 3.1). The decreases in total mortality rates were greater for the white than the colored populations and both white and colored women had lower mortality rates than their male counterparts. Decreases occurred for both old and young age groups. For example, considering policyholders ages 65-74, annualized average total mortality rates per 1000 policyholders decreased between 1911-5 and 1931-35 from 83.1 to 67.5 for white men, from 69.3 to 53.8 for white women, from 86.1 to 77.8 for colored men, and from 72.9 to 67.7 for colored women. For those ages 35-44, the decreases for white men were from 16.4 to 7.7, for white women from 9.6 to 5.2, for colored men from 19.2 to 14.4, and for colored women from 16.6 to 12.1.

Mortality rates from angina pectoris, the diagnostic term used for coronary heart disease, for white and colored men and women in 1911-15 indicate that it was an unimportant cause of death for both sexes that was somewhat more prevalent in older age groups (Table 3.1). Annualized average angina pectoris mortality rates for each race-sex group ages 35-44 and 45-54 constituted no more than 0.4 percent of total mortality and for each group ages 55-64 and 65-74 no more than 0.9 percent of total mortality. The increases in mortality rates with age were consistent with the hardening of the arteries theory. Sex differences were small and inconsistent except that white men had slightly higher mortality rates than the other three groups, primarily at ages 65-74.

Table 3.1 Mortality Rates from Coronary Heart Disease and All Causes by Age, Race, and Sex for Metropolitan Industrial Life Insurance Policyholders, 1911-1935 (Annualized average or annual rates per 1,000 policyholders)

						All Causes			
	WM	*WF*	*CM*	*CF*		*WM*	*WF*	*CM*	*CF*
AGES 35-44									
Angina pectoris (five-year annualized average)									
1911-15	0.04	0.02	0.07	0.08		16.4	9.6	19.2	16.6
1931-35	0.12	0.02	0.12	0.08		7.7	5.2	14.4	12.1
Diseases of coronary arteries									
1930	0.06	0.02	0.07	0.03					
1935	0.27	0.06	0.14	0.13					
AGES 45-54									
Angina pectoris (five-year annualized average)									
1911-15	0.12	0.05	0.10	0.18		24.3	15.0	28.2	24.5
1931-35	0.42	0.10	0.25	0.18		16.0	10.3	24.4	20.0
Diseases of coronary arteries									
1930	0.20	0.07	0.08	0.08					
1935	0.86	0.23	0.46	0.26					
AGES 55-64									
Angina pectoris (five-year annualized average)									
1911-15	0.30	0.18	0.24	0.23		42.0	30.8	49.0	41.3
1931-35	0.92	0.33	0.42	0.33		33.0	23.0	42.4	35.4
Diseases of coronary arteries									
1930	0.48	0.20	0.07	0.10					
1935	1.89	0.79	0.84	0.61					
AGES 65-74									
Angina pectoris (five-year annualized average)									
1911-15	0.73	0.41	0.37	0.35		83.1	69.3	86.1	72.9
1931-35	1.40	0.78	0.60	0.56		67.5	53.8	77.8	67.7
Diseases of coronary arteries									
1930	0.68	0.38	0.33	0.22					
1935	3.14	1.67	1.19	1.01					

WM = white male, WF = white female, CM = colored male, CF = colored female

Sources: Louis I. Dublin and Alfred J. Lotka, *Twenty-five Years of Health Progress* (New York: Metropolitan Life Insurance Company, 1937), pp. 16, 277, 280

An analysis of the increase in coronary heart disease mortality rates in the 1930s requires the use of two disease categories: the original "angina pectoris" category plus the additional new diagnostic category introduced in 1930, "diseases of the coronary arteries." A comparison of the mortality rates in the two 1930s categories provides strong evidence that physicians chose one or the other category as a matter of habit or personal preference rather than differences in the types of the diseases. The differences in mortality rates by sex and race within each age group are practically the same for angina pectoris in 1931-35 and for diseases of the coronary arteries in 1930 and 1935 (Table 3.1). This makes it appropriate to compare the 1911-15 annualized average mortality rates for angina pectoris to the combination of the 1931-35 annualized average mortality rates for angina pectoris and the 1935 mortality rates for diseases of the coronary arteries.

Mortality rates must be analyzed separately for each age group because experts considered the increase in coronary heart disease mortality rates to be primarily a consequence of the greater longevity of the population, consistent with the hardening of the arteries theory. Dublin and Lotka stated in 1937 that "the startling rise in coronary disease mortality is not really a cause for alarm" because it was due to more accurate diagnosis and a larger number of older persons.[8]

The error of this conclusion is shown by the substantial increases in coronary heart disease mortality rates from 1911-15 to 1931-35 for most age groups from 35-44 to 65-74 for both sexes and both races (Table 3.1). Considering only angina pectoris, annualized average mortality rates increased from 1911-15 to 1931-35 for both white and colored men in all age groups and for both white and colored women in age groups 55-64 and 65-74. Considering only diseases of the coronary arteries, mortality rates increased substantially in the five year period from 1930 to 1935 for all age groups of each of the four race-sex groups.

A more useful measure of the increases in mortality rates is to compare annualized average mortality rates from angina pectoris in 1911-15 to the combination of annualized average mortality rates from angina pectoris in 1931-35 plus mortality rates from diseases of the coronary arteries in 1935 (Table 3.1). All age groups and both sexes experienced increases in mortality rates but the increases were greater for older age groups, for men than women in both races, and for the older white than the older colored population. For example, among those ages 35-44, the mortality rates per 1000 increased from 1911-15 to 1931-35 for white men from 0.04 to 0.39, for white women from 0.02 to 0.08, for colored men from 0.07 to 0.26, and for colored women from 0.08 to 0.21. At ages 65-74, the mortality rates per 1000 increased from 1911-15 to 1931-35 for white men from 0.73 to 4.54, for white women from 0.41 to 2.45, for colored men from 0.37 to 1.79, and for colored women from 0.35 to 1.57.

These increases are in striking contrast to the improvements in the general health of the groups as measured by the decreases in their annualized average total mortality rates excluding coronary heart disease (Table 3.1). The excluded diseases are angina pectoris in 1911-15 and the combination of angina pectoris in 1931-35 and diseases of the coronary arteries in 1935. All age groups of both sexes experienced decreases but the older groups, who had greater increases in their coronary heart disease mortality rates, had greater decreases in their total mortality rates excluding coronary heart disease. For example, for those ages 65-74, between 1911-15 and 1931-35 annualized average total mortality rates excluding coronary heart disease per 1000 persons decreased for white men from 82.4 to 63.0, for white women from 68.9 to 51.4, for colored men from 85.7 to 76.0, and for colored women from 72.6 to 66.1. At ages 35-44, the decreases for white men were from 16.4 to 7.3, for white women from 9.6 to 5.1, for colored men from 19.1 to 14.1, and for colored women from 16.5 to 11.9.

This analysis has shown that coronary heart disease in the 1930s was a different type of coronary heart disease than occurred in 1911-15. All age groups of men and women of both races experienced increases in their mortality rates, demonstrating that the increases were not due to more elderly persons in the population and could not be explained by hardening of the arteries. The increases were much greater for men than women, which demonstrated that the new coronary heart disease differed in major respects from the disease early in the century. Mortality rates excluding coronary heart disease decreased in all age, sex, and race groups, demonstrating that the increases were not caused by poorer health of the population. The oldest age groups had both the greatest increases in coronary heart disease mortality rates and the greatest decreases in mortality rates from all other causes.

This new type of coronary heart disease produced the highest mortality rates in white men of the four race-sex groups for all age groups (Table 3.1). White women had the lowest mortality rates of the four race-sex groups at ages 35-44 and 45-54 but the second highest rates at older ages. Considering mortality rates for the combination of angina pectoris in 1931-35 and diseases of the coronary arteries in 1935, at ages 45-54 the annualized average mortality rates per 1000 persons were highest for white men (1.28) with decreasingly lower rates for colored men (0.71), colored women (0.44), and white women (0.33). At ages 65-74 the highest mortality rates continued to be for white men (4.54), now followed by white women (2.45), colored men (1.79), and colored women (1.57).

The substantial increases in coronary heart disease mortality rates from 1911-15 to 1931-35 among Metropolitan Life Insurance Company policyholders were not matched by increases in other major diseases that affected the heart and arteries (Table 3.2). Sex and race differences in mortality rates from these other diseases also differed substantially from those for coronary heart disease. Stroke is particularly relevant because the great majority of strokes consist of a blockage of an artery in the brain. Considering mortality rates for Metropolitan Life Insurance Company industrial policyholders ages 45-74, annualized average age-adjusted mortality rates from stroke or cerebral hemorrhage (including paralysis) as well as for chronic nephritis (kidney disease) decreased for all of the four race-sex groups from 1911-15 to 1931-35. For example, stroke mortality rates per 1000 for white men decreased from 4.0 to 2.5, for white women from 3.6 to 2.3, for colored men from 4.7 to 4.2, and for colored women from 5.4 to 4.6. Annualized average mortality rates for both organic heart diseases and diabetes mellitus changed very slightly. Trends in hypertensive disease mortality rates were not available, but hypertension is a major risk factor for stroke and the decrease in stroke mortality rates makes an increase in hypertensive disease mortality rates highly unlikely.

Table 3.2 Mortality Rates from Selected Diseases by Race and Sex, Metropolitan Industrial Life Insurance Policyholders Ages 45-74, 1911-1935
(Age adjusted annualized average mortality rates per 1,000 policyholders)

	1911-15				**1931-35**			
	WM	*WF*	*CM*	*CF*	*WM*	*WF*	*CM*	*CF*
Cerebral hemorrhage	4.0	3.6	4.7	5.4	2.5	2.3	4.2	4.6
Chronic nephritis	5.3	3.6	6.8	4.6	2.7	2.3	5.3	4.4
Organic heart diseases	6.6	5.6	8.4	7.6	7.1	5.2	9.4	8.1
Diabetes mellitus	0.5	0.8	0.4	0.4	0.6	1.3	0.5	1.2

WM = white male, WF = white female, CM = colored male, CF = colored female

Source: Louis I. Dublin and Alfred J. Lotka, *Twenty-five Years of Health Progress* (New York: Metropolitan Life Insurance Company, 1937), pp. 251, 288, 302, 320.

Thus coronary heart disease in the early 1930s was a different type of disease than the same disease about 1910. Mortality rates increased for younger age groups, which could not be explained by the hardening of the arteries theory. Mortality rates increased by larger amounts for older persons, despite the improvements in their general health that would reduce the probability of hardening of the arteries. Much greater increases in mortality rates occurred for men than women, which differed from the disease early in the century. The accuracy of the diagnoses of the new coronary heart disease is clearly indicated by the widening sex differences in mortality rates that did not occur with other relevant diseases. No relationship existed between the coronary heart disease mortality rates of each of the four groups and its mortality rates from all other causes. Mortality rates from other diseases of the heart and arteries did not increase, providing indisputable evidence that the increases in coronary heart disease rates were not part of a general pattern of increases in diseases affecting these organs.

These data from 1911 to the 1930s demonstrate that a new form of coronary heart disease appeared that produced rapid increases in mortality rates and differed in key respects from coronary heart disease early in the century. Its characteristics were similar to those of generally accepted pandemics described in Chapter 1. The data were based on the experiences of a reasonably homogeneous population of millions of urban adults who owned inexpensive life insurance policies sold by the Metropolitan Life Insurance Company. Beginning in 1940 federal vital statistics were available to study the peak of the coronary heart disease pandemic that occurred in the United States in the following decades.

References

1. A history of the emergence of the coronary heart disease pandemic is provided in William G. Rothstein, *Public Health and the Risk Factor: A History of an Uneven Medical Revolution* (Rochester, NY: University of Rochester Press, 2003), pp. 182-217.
2. Robert I. Levy, Howard G. Bruenn, and Dorothy Kurtz, "Facts on Disease of the Coronary Arteries, Based on a Survey of the Clinical and Pathological Records of 762 Cases," *American Journal of the Medical Sciences* 187 (1934): 376-90; T.W. Anderson and W.H. LaRiche, "Ischemic Heart Disease and Sudden Death, 1901-1961," *British Journal of Preventive and Social Medicine* 24 (1970): 1-9.
3. Robert D. Grove and Alice M Hetzel, *Vital Statistics Rates in the United States, 1940-1960* (Washington, DC: National Center for Health Statistics, 1968), pp. 7-9.
4. Rothstein, *Public Health and the Risk Factor*, pp. 55-61.
5. Louis J. Dublin and Alfred J. Lotka, *Twenty-Five Years of Health Progress: A Study of the Mortality Experience among the Industrial Policyholders of the Metropolitan Life Insurance Company 1911 to 1935* (New York: Metropolitan Life Insurance Company, 1937), pp. 7-10; U.S. Bureau of the Census, *Historical Statistics of the United States, Colonial times to 1970*. 2 vols. (Washington, DC: 1975), p. I:8.
6. I.S. Falk, "Louis I. Dublin: November 1, 1882–March 7, 1969," *American Journal of Public Health* 59 (1969): 1083-85; Louis I. Dublin, "Alfred James Lotka, 1880-1949," *Journal of the American Statistical Association* 45 (1950): 138-39; Rothstein, *Public Health and the Risk Factor*, p. 174.
7. Dublin and Lotka, *Twenty-Five Years of Health Progress*, pp. 15, 27.
8. Dublin and Lotka, *Twenty-Five Years of Health Progress*, p. 284.

Chapter 4

The Coronary Heart Disease Pandemic in the United States from 1940 to 1967

Coronary heart disease mortality became the leading cause of adult deaths in the United States from the 1940s to near the end of the century. Its distinctive characteristics demonstrate that this was a new pandemic coronary heart disease that differed significantly from normal coronary heart disease early in the century. Mortality rates increased substantially and rapidly for all age and sex groups of the white and black populations. Greater increases in mortality rates occurred among older than younger persons and among men than women in each race. The pandemic produced higher mortality rates in urban than rural populations and in lower than higher socioeconomic groups. Federal government vital statistics were used to measure coronary heart disease mortality rates beginning in 1940.

To summarize the increases from 1940 to 1967, coronary heart disease mortality rates per 1000 at ages 45-54 increased for white men from 1.7 to 3.4, for white women from 0.4 to 0.7, for black men from 1.1 to 3.2, and for black women from 0.8 to 1.6. At ages 65-74 they increased for white men from 6.7 to 19.3, for white women from 3.4 to 8.8, for black men from 2.0 to 14.3, and for black women from 1.4 to 9.3.

The Coronary Heart Disease Pandemic in 1940

Many pandemics exhibit their most distinctive population characteristics after they emerge and before they diffuse throughout the affected populations. For example, the influenza pandemic emerged in population groups with personal characteristics that differed strikingly from the persons who normally develop the disease. The lung cancer pandemic first occurred in men who were never exposed to the recognized carcinogens. The distinctiveness of these groups made health professionals realize the novel characteristics of the outbreaks of these two diseases. For this reason a description of coronary heart disease during its early period in 1940 is valuable, especially because of the availability of federal government mortality statistics. This was less than a decade after the disease became a significant health problem based on Metropolitan Life Insurance Company mortality statistics.

United States vital statistics in 1940 used the same category, "angina pectoris and diseases of the coronary arteries" as the Metropolitan Life Insurance Company statistics of the 1930s. The term coronary heart disease will be used for brevity. A comparison of the federal vital statistics mortality rates in 1940 and the Metropolitan Life Insurance Company mortality rates in the early 1930s shows many similarities in mortality rates by age, sex, and race. This provides strong evidence of the representativeness of the Metropolitan Life Insurance Company policyholders and the usefulness of the data as an indicator of the emergence of the pandemic.

The federal vital statistics available to investigate the population groups affected by the pandemic in the United States from 1940 to 1960 are extraordinarily thorough and detailed. In 1943 the United States Bureau of the Census published a historical volume on vital statistics in the United States from 1900 to 1940 and in 1968 the National Center for Health Statistics published a continuation volume for the 1940-60 period. By 1940 all 48 states then in the union were included in the death registration area. Both volumes used new and innovative methods of statistical analysis and contain numerous detailed reports on specific causes of death in population groups and geographic areas. They are remarkable achievements.[1]

While the Metropolitan Life Insurance Company data used "colored" as a racial category, the authors of the 1900-40 report used the more inclusive terms "all other races" or "nonwhite," which included blacks, American Indians, Asian nationalities, and others. The authors stated that "the classification of mortality and nativity by race is difficult, and the results tend to be ambiguous and incomplete." Differences in mortality are mostly likely due to different "economic, social, and medical circumstances" of the groups rather than biological differences, for which "very little evidence" exists. They stated: "An observed difference in mortality between races may in actuality be no more than a difference of mortality for different economic classes." Nationally, in 1940 the nonwhite population constituted 10 percent of the 18.3 million persons ages 35-44, 8 percent of the 15.5 million persons ages 45-54, 7 percent of the 10.6 million persons ages 55-64, and 7 percent of the 6.4 million persons ages 65-74. In 1940, 95 percent of the total nonwhite population was black so that the term black will be used to describe the national mortality rates of nonwhites. The 1940-60 volume also used the category "nonwhite."[2]

The new pandemic form of coronary heart disease became a more important cause of death in all age groups in 1940 (Table 8.1). Among white men deaths from the category of "angina pectoris and diseases of the coronary arteries" accounted for between 9 and 15 percent of all deaths for each age group from 35-44 to 65-74. Among white women the category was responsible for 3 percent of total mortality at age 35-44 rising to 8 percent at ages 65-74. For black men and women, the proportion of deaths from the category in each group was no more than 5 percent in all age groups, but these low percentages were due in part to their higher mortality rates from all other causes. For this reason, this analysis will focus on coronary heart disease mortality rates rather than the proportion of all deaths that were caused by coronary heart disease.

One of the most striking aspects of this new form of coronary heart disease in 1940 was the large differences in mortality rates among age, sex, and race groups, something that did not exist in 1911-15 (Tables 3.1 and 4.1). In 1940 white men of all ages and older white women had higher mortality rates than their black counterparts despite their lower mortality rates from all other causes. In every age group white men had the highest coronary heart disease mortality rates per 1000 persons, and their rates increased from 0.5 at ages 35-44 to 6.7 at ages 65-74. White women had the lowest mortality rates of the four race-sex groups at younger ages but the second highest rate at ages 65-74. Their rates increased from 0.1 at ages 35-44 to 3.4 at ages 65-74. Black men had mortality rates equal to those of white men at younger ages but much lower mortality rates at older ages. Their rates increased from 0.5 at ages 35-44 to 2.0 at ages 65-74. Black women had slightly higher rates than white women at younger ages but the lowest rates of the four groups at the oldest ages. Their rates increased from 0.3 at ages 35-44 to 1.4 at ages 65-74.

The large sex differences in coronary heart disease mortality rates in 1940 were greater in the white population and widened more with age. For example, white men had 0.4 more deaths per 1000 persons than white women at ages 35-44 and 4.3 more deaths at ages 65-74. Black men had 0.2 more deaths than black women at ages 35-44 and 0.6 more deaths at ages 65-74.

Despite the higher coronary heart disease mortality rates of white than black men in all age groups and of white than black women in the oldest age groups, black men and women had higher mortality rates per 1000 persons from all other causes at all ages in 1940. For example, at ages 65-74 the coronary heart disease mortality rate of white men was 6.7 compared to 2.0 for black men, but their mortality rate from all other causes was 47.3 compared to 54.5. The situation was similar for the women. At ages 65-74 white women had coronary heart disease mortality rates of 3.4 compared to 1.4 for black women, but mortality rates from all other causes of 38.1 compared to 44.9 for black women.

In 1940 coronary heart disease mortality rates were much higher in urban than rural areas, indicating that urban areas provided a more receptive environment for the new pandemic disease (Table 4.1). Both coronary heart disease and total mortality rates for ages 35-44 to 65-74 in each of the four race-sex groups were similar in towns and cities that varied from 2,500-10,000 persons to more than 100,000 persons. In rural areas, however, mortality rates from angina pectoris and diseases of the coronary arteries were between 65 and 80 percent of those in towns and cities for age groups from 45-54 to 65-74 of each of the four race-sex groups. The same patterns existed to a lesser degree for total mortality excluding coronary heart disease, with rural areas having lower total mortality rates in all age, sex, and race groups. The lower mortality rates in rural areas benefitted almost half of the population. In 1940, 42 percent of the white and 52 percent of the nonwhite population lived in rural areas and 9 and 7 percent respectively lived in towns of 2,500-10,000 population.[3]

Table 4.1 United States Mortality Rates from Coronary Heart Disease and All Causes by Age, Race, Sex, and City Population, 1940 (Rates per 1,000 persons)

	Angina Pectoris and Diseases of Coronary Arteries				*All Causes*			
	WM	*WF*	*MO*	*FO*	*WM*	*WF*	*MO*	*FO*
Ages 35-44								
United States	0.5	0.1	0.5	0.3	5.1	3.7	13.2	11.7
City population								
100,000 or more	0.6	0.1	0.5	0.3	5.6	3.8	14.1	11.7
10,000-100,000	0.6	0.1	0.6	0.5	5.2	3.7	15.5	13.2
2,500-10,000	0.6	0.1	0.6	0.6	5.9	4.0	17.1	14.1
Rural	0.3	0.1	0.3	0.3	4.4	3.4	10.9	10.7
Ages 45-54								
United States	1.7	0.4	1.1	0.8	11.4	7.5	24.5	21.1
City population								
100,000 or more	2.0	0.4	1.4	0.9	13.5	8.3	27.4	22.6
10,000-100,000	2.1	0.4	1.5	1.0	12.1	7.8	29.3	24.7
2,500-10,000	1.9	0.4	1.5	0.9	12.6	7.8	31.0	24.3
Rural	1.1	0.3	0.7	0.5	9.0	6.4	20.0	18.1
Ages 55-64								
United States	3.8	1.2	1.8	1.2	25.2	16.8	39.5	35.7
City population								
100,000 or more	4.6	1.5	2.2	1.4	30.3	19.2	44.8	37.9
10,000-100,000	4.9	1.4	2.7	1.6	27.6	17.6	50.5	43.2
2,500-10,000	4.4	1.3	2.6	1.5	27.9	17.1	52.7	42.8
Rural	2.7	1.0	1.2	0.9	20.0	14.4	32.1	30.9
Ages 65-74								
United States	6.7	3.4	2.0	1.4	54.0	41.5	56.5	46.3
City population								
100,000 or more	7.7	3.6	2.9	2.0	62.4	45.7	68.4	52.8
10,000-100,000	8.4	3.9	2.8	1.6	59.3	42.3	65.7	49.1
2,500-10,000	8.2	3.7	3.2	1.4	59.4	41.6	65.5	49.2
Rural	5.2	2.7	1.3	1.1	46.2	37.8	49.3	41.7

WM = white male: WF = white female, MO = male other races, FO =female other races

Sources: Forrest L. Linder and Robert D. Grove, *Vital Statistics Rates in the United States, 1900-1940* (Washington, DC: United States Government Printing Office, 1943), pp. 534-53.

The lower mortality rates of rural residents cannot be explained by different diagnostic criteria or methods of medical care in rural areas. Very few physicians practiced in rural areas so rural patients went to nearby towns or cities for medical care.[4] Total and coronary heart disease mortality rates were practically the same in towns of 2,500-10,000 population as in cities of more than 100,000 population.

The Pandemic from 1950 to 1967

Arteriosclerotic heart disease became the major source of mortality at mid-century. Its importance as a cause of death varied substantially by age, race, and sex (Table 8.1). In 1960 arteriosclerotic heart disease in white men was responsible for 26 percent of all deaths ages 35-44 and about 40 percent of all deaths at ages 45-54, 55-64, and 65-74. Among white women it was responsible for only 7 percent of deaths ages at 35-44 but this increased steadily to 26 percent at ages 55-64 and 33 percent at ages 65-74. The proportions were lower among black men and women, mostly because of their higher mortality rates from other causes. Among black men it was responsible for 12 percent of all deaths at ages 35-44 and increased to about 24 percent of those ages 55-64 and 65-74. Among black women the percentages increased from 8 percent at ages 35-44 to 20 percent at ages 55-64 and 23 percent at ages 65-74.

The peak of the coronary heart disease pandemic in the United States occurred in the 1950s and 1960s, after which mortality rates decreased steadily, as will be shown in later chapters. Many characteristics of the pandemic remained unchanged during the peak period but, as occurs with most pandemics, it spread to population groups that were less affected at its onset.

In enumerating coronary heart disease mortality rates from 1948 to 1967, the United States vital statistics reports used the new term of the International Classification of Diseases, "arteriosclerotic heart disease, including coronary disease." This new term was adopted in the sixth revision (1948-57) and continued to be used in the seventh revision (1958-67). It combined two terms that had been listed in different categories in the fifth revision: "angina pectoris and disease of the coronary arteries" and "arteriosclerotic heart disease."[5]

The period from 1950 to 1967 was one of improvements in the health of adults of all ages in the United States (Table 4.2). Decreases in total mortality rates excluding arteriosclerotic heart disease occurred for white and black men and women in all age groups except for black men ages 65-74. For example, from 1950 to 1967 at ages 55-64, total mortality rates excluding arteriosclerotic heart disease per 1000 decreased from 14.9 to 13.3 for white men, from 10.2 to 7.6 for white women, from 29.3 to 23.4 for black men, and from 23.9 to 16.3 for black women.

The coronary heart disease pandemic reached its peak from about 1950 to about 1967 using arteriosclerotic heart disease, with the timing varying among younger and older age groups (Table 4.2). The peak period began about 1950 for younger age groups and changed little to 1967. For example, at ages 45-54 arteriosclerotic heart disease mortality rates per 1000 persons for white men increased from 3.2 in 1950 to 3.4 in 1967 while those for black men increased from 2.5 in 1950 to 3.2 in 1967. White women ages 45-54 had mortality rates per 1000 of 0.7 in both years and black women ages 45-54 had rates of 1.7 in 1950 and 1.6 in 1967.

Table 4.2 United States Mortality Rates from Arteriosclerotic Heart Disease and All Causes by Age, Sex, and Race, 1950-1967
(Rates per 1,000 persons)

Age	*Arteriosclerotic Heart Disease*				*All Causes*			
	WM	*WF*	*NM*	*NF*	*WM*	*WF*	*NM*	*NF*
35-44								
1950	0.8	0.1	0.7	0.5	3.8	2.4	8.6	7.5
1960	0.9	0.1	0.9	0.5	3.3	1.9	7.3	5.5
1967	0.9	0.1	1.0	0.5	3.4	1.9	8.5	5.1
45-54								
1950	3.2	0.7	2.5	1.7	9.8	5.5	18.6	15.5
1960	3.5	0.6	3.0	1.7	9.3	4.6	15.5	11.5
1967	3.4	0.7	3.2	1.6	9.0	4.6	16.1	10.1
55-64								
1950	8.1	2.7	5.5	3.7	23.0	12.9	34.8	27.6
1960	9.0	2.8	7.2	4.8	22.3	10.8	31.5	24.1
1967	8.9	2.5	7.5	4.3	22.2	10.1	30.9	20.6
65-74								
1950	16.1	8.4	9.4	6.6	48.6	32.4	57.9	46.1
1960	19.1	9.2	13.4	9.0	48.5	27.8	56.6	39.8
1967	19.3	8.8	14.3	9.3	48.9	25.9	65.0	44.1

WM = white male: WF = white female, NM = nonwhite male, NF = nonwhite female

1950-1960 mortality: Robert D. Grove and Alice M. Hetzel, *Vital Statistics Rates in the United States, 1940-1960* (Washington, DC: National Center for Health Statistics, 1968), pp. 376-78, 457-59.

1967 arteriosclerotic heart disease mortality: Millicent W. Higgins and Russell V. Luepker, eds., *Trends in Coronary Heart Disease Mortality: The Influence of Medical Care* (New York: Oxford University Press, 1988), pp. 284-87.

1967 total mortality: United States Public Health Service, *Vital Statistics of the United States 1968, Vol. II-Mortality Part A* (Washington, DC: U.S. Government Printing Office, 1972), pp. 1-4, 1-5 [http://www.cdc.gov/nchs/data/vsus/mort68_2a.pdf] (Accessed Feb. 27, 2015).

By contrast, most groups of older persons experienced steadily increasing arteriosclerotic heart disease mortality rates per 1000 persons from 1950 to 1967. At ages 65-74 the mortality rates per 1000 for white men increased from 16.1 in 1950 to 19.3 in 1967 and for black men from 9.4 to 14.3. Mortality rates per 1000 for white women ages 65-74 increased slightly from 8.4 in 1950 to 8.8 in 1967. Mortality rates for black women ages 65-74 increased to a greater extent and became slightly higher than those of white women, rising from 6.6 in 1950 to 9.3 in 1967.

The increases in mortality rates among older age groups and the lack of increases among younger age groups between 1950 and 1967 widened the age differences in arteriosclerotic heart disease mortality rates, primarily among men. For example, white men ages 65-74 had 12.9 more deaths per 1000 than those ages 45-54 in 1950 but 15.9 more deaths in 1967. The difference between white women in the two age groups increased slightly from 7.7 more to 8.1 more deaths. The increases in the mortality rate differences between those ages 65-74 and 45-54 in the black population were greater, from 6.9 to 11.3 more deaths for older black men and from 4.9 to 7.7 more deaths for older black women. Age differences in mortality rates from all other causes narrowed for the white population but widened for the black population from 1950 to 1967.

The sex differences in arteriosclerotic heart disease mortality rates per 1000 persons widened from 1950 to 1967, mostly for older age groups. At ages 65-74 the sex difference for the white population increased from 7.7 more deaths for men in 1950 to 10.5 more deaths in 1967 and for the black population from 2.8 more deaths for men to 5.0 more deaths. At ages 45-54, the increases were from 2.5 to 2.7 deaths more deaths for men in the white population and from 0.8 to 1.6 more deaths for men in the black population. Sex differences in total mortality rates excluding arteriosclerotic heart disease widened for the white population at older age groups and for the black population in all age groups.

As the pandemic progressed, it spread to the black population and narrowed the differences in mortality rates between the white and black groups, primarily at older ages. For example, at ages 65-74 white men had 6.7 more deaths per 1000 population than black men in 1950 but only 5.0 more deaths in 1967. Mortality rates among white women stabilized over the period but those of black women continued to increase. White women ages 65-74 had 1.8 more deaths than black women in 1950 but 0.5 fewer deaths in 1967.

The pandemic also spread from urban to rural populations as shown by the narrowing of urban-rural differences in coronary heart disease mortality rates. In 1940 mortality rates were 20-35 percent lower in rural than urban areas (Table 4.2). In 1950 the Census Bureau used a new system of classification in which a metropolitan area was defined as one or more counties around a center city of at least 50,000 population. Using this definition, in 1959-61 arteriosclerotic heart disease mortality rates were almost always less than 20 percent lower in nonmetropolitan counties than in metropolitan areas with center cities. This occurred for white and nonwhite men and women in all of the relevant age groups.[6]

Coronary heart disease mortality rates can be compared to other diseases that also involve arteries to demonstrate the highly atypical sex differences in arteriosclerotic heart disease mortality rates (Table 4.3). Age adjusted arteriosclerotic heart disease mortality rates in 1960 were higher for white men than white women and black men than black women (0.7 more deaths per 1000 in each case) while the sex differences in mortality rates for hypertensive disease and stroke were insignificant. This provides further evidence that the increase in sex differences in arteriosclerotic heart disease at the peak of the pandemic resulted from the pandemic. It also demonstrates the ability of physicians to differentiate arteriosclerotic heart disease from other relevant diseases.

Table 4.3 United States Age-Adjusted Mortality Rates for Selected Artery-Related Diseases, 1960
(Rates per 1,000 persons)

	White Male	*White Female*	*Nonwhite Male*	*Nonwhite Female*
Arteriosclerotic heart disease	3.1	1.4	2.2	1.5
Hypertensive disease	0.3	0.3	0.9	0.9
Stroke	0.8	0.7	1.4	1.3

Source: Robert D. Grove and Alice M. Hetzel, *Vital Statistics Rates in the United States, 1940-1960* (Washington, DC: National Center for Health Statistics, 1968), pp. 369-71.

Coronary Heart Disease and Socioeconomic Position

The relationship between coronary heart disease and socioeconomic position became of concern at the emergence of the pandemic in the 1930s. Persons of high socioeconomic positions obtained their medical care from prominent physicians, who reported their experiences to the medical profession and the public. Most persons of low socioeconomic position did not have personal physicians and those who did were treated by physicians who seldom reported their experiences. The consequence was the erroneous conclusion from the observation of small numbers of atypical patients early in the pandemic that persons of higher socioeconomic positions had higher rates of coronary heart disease than those of lower socioeconomic positions.

Higher coronary heart disease rates in persons of lower socioeconomic positions were found in three methodologically sophisticated studies of very large samples. Two measured mortality rates, one using a sample of millions of purchasers of Metropolitan Life Insurance Company life insurance policies and the other a sample of many thousands of army veterans who received life insurance policies issued by the army. Reporting of death was essentially complete in the two studies because payment of life insurance death benefits required a death certificate. The third study of employees of a large private corporation measured new cases of myocardial infarctions of employees that were reported to the corporation by its health insurance company. Reporting in this study was also essentially complete.

The methodologies of the studies had many features that were rarely used in other studies. The Metropolitan Life Insurance Company and army veterans studies used samples that were many times larger than most other studies and had essentially complete reporting of deaths. All three studies followed their samples for years to determine which persons did and did not develop or die of coronary heart disease. The samples of all three studies consisted of persons healthy enough to obtain life insurance, serve in the armed forces, or be employed. This reduces the effect of other health problems on the members of the samples. The participants in all three studies were typical of large elements of the United States population and included a wide range of socioeconomic groups. The Metropolitan Life Insurance Company study included only white policyholders and the other two studies consisted predominantly of white persons.

The studies differed in the outcomes that they measured. The Metropolitan Life Insurance Company and the army veteran studies used mortality rates, while the private corporation study used all new cases of myocardial infarctions. This permits the comparison of two studies of persons who died of coronary heart disease with a study of all persons who developed a major form of the disease regardless of outcome.

The Metropolitan Life Insurance Company study compared mortality rates for 1935-39 between its millions of low income urban white policyholders with inexpensive industrial life insurance policies and its millions of middle and upper income white policyholders with much more expensive ordinary life insurance policies. Black policyholders were excluded because very few had ordinary life insurance. The listed cause of death was diseases of the heart and arteries, but the large sex difference indicated that coronary heart disease was the major component of the category. This conclusion is supported by the lack of sex differences in mortality rates from stroke and chronic nephritis. For those ages 36-45 heart disease mortality rates per 1000 policyholders were 1.3 for men and 0.7 for women among industrial policyholders and 0.8 for men and 0.4 for women among ordinarily policyholders. For ages 46-55 heart disease mortality rates per 1000 were 4.1 for men and 2.0 for women among industrial policyholders and 2.8 for men and 1.2 for women among ordinary policyholders. Among policyholders ages 56-65 heart disease mortality rates were 10.5 for men and 5.9 for women among industrial policyholders and 7.8 for men and 4.1 for women among ordinary policyholders.[7]

Industrial policyholders also had higher total mortality rates excluding heart disease than ordinary policyholders, which supports the validity of the data. For ages 36-45 the mortality rates per 1000 policyholders were 5.3 for male and 3.7 for female industrial policyholders and 3.0 for male and 2.9 for female ordinary policyholders. At ages 46-55 the mortality rates were 10.3 for male and 7.0 for female industrial policyholders and 6.4 for male and 5.4 for female ordinary policyholders. At ages 56-65 the mortality rates were 20.4 for male and 15.0 for female industrial policyholders and 13.8 for male and 10.6 for female ordinary policyholders.[8]

A second study followed 85,000 white male U.S. army veterans from their discharge in 1946 to 1969 and found higher coronary heart disease mortality rates in veterans who had lower military ranks when in the service. The average age at discharge was 24 years and the average age at the termination of the study was 46 years. The members of the sample were healthier than average, as indicated by an overall mortality rate during the course of the study that was 84 percent of the rate of the comparable United States population. The coronary heart disease mortality rate of commissioned officers was 50 percent of the general population of the same age distribution, that of noncommissioned officers 88 percent, and that of privates 98 percent. Similar differences in total mortality rates occurred for the three groups. Veterans in each rank with more education had lower total mortality rates than those with less education.[9]

Another study examined 1331 new cases of myocardial infarctions among 73,573 male employees ages 17-64 of a very large industrial corporation, E.I. DuPont de Nemours and Company, from 1956 to 1961. It found lower rates of myocardial infarctions in employees with higher position in the company. All members of the sample were healthy enough to be employed full-time, worked for the same company, and had the same health insurance benefits. The listed occupations were those used by the company and were much more narrowly defined and internally consistent than census categories. For these reasons they represented meaningful socioeconomic differences among the groups. Annualized age-adjusted myocardial infarction rates per 1000 employees of between 2.2 and 2.5 occurred among salaried workers, including managers, professionals, and salesmen, Rates between 3.5 and 4.0 occurred among lower paid hourly wage employees and first-level supervisors. Studies of samples of workers in the different occupational groups found that differences among them in blood pressure levels, serum cholesterol levels, smoking rates, and body weight were too small to have caused the differences in disease rates.[10]

These findings based on long-term studies of very large samples of three different population groups permit several conclusions. Coronary heart disease rates in all three studies were greater among lower than higher socioeconomic groups. The members of the samples were all healthy enough to be employed at the beginning of the studies, which reduced the likelihood that poorer health produced the higher rates in the lower socioeconomic groups. Lower socioeconomic groups also had higher total mortality rates excluding coronary heart disease, a universally observed finding that supports the validity of the studies. The two studies that used mortality rates as the outcome produced findings that were very similar to the study that used all new cases of myocardial infarctions, which included both survivors and deceased. This indicates that studies using mortality rates are a valid measure of overall trends in the disease.

Thus substantial evidence exists that from the 1930s through the 1960s a new form of coronary heart disease emerged and became the most significant cause of mortality among middle aged and older adults. Mortality rates increased much more among certain population groups that did not have higher rates of the disease before the pandemic early in the century. High mortality rates occurred first in urban areas and then spread to rural areas. Diagnostic misclassifications were not a significant issue.

The emergence of coronary heart disease as a major cause of mortality aroused great concern among experts and the general population. Particular attention was devoted to identifying those causes of the disease that could be used for prevention because of the severity of the disease and the paucity of available treatments. The research on prevention produced a number of risk factors that will be examined and evaluated in the next chapter.

References

1. Forrest E. Linder and Robert D. Grove, *Vital Statistics Rates in the United States, 1900-1940* (Washington, DC: Bureau of the Census, 1943); Robert D. Grove and Alice M. Hetzel, *Vital Statistics Rates in the United States, 1940-1960* (Washington, DC: National Center for Health Statistics, 1968).
2. Linder and Grove, *Vital Statistics Rates in the United States, 1900-1940*, pp. 12, 872; U.S. Bureau of the Census, *Historical Statistics of the United States. Colonial Times to 1970* 2 vols. (Washington, DC: 1975), p. I:14.
3. Linder and Grove, *Vital Statistics Rates in the United States, 1900-1940*, pp. 934-5.
4. William G. Rothstein, *American Medical Schools and the Practice of Medicine: A History* (New York; Oxford University Press, 1987), pp. 119-20.
5. See chapter 3.
6. Iwao M. Moriyama, Dean E. Krueger, and Jeremiah Stamler, *Cardiovascular Diseases in the United States* (Cambridge, MA: Harvard University Press, 1971), pp. 24, 77.
7. William G. Rothstein, *Public Health and the Risk Factor: A History of an Uneven Medical Revolution* (Rochester, NY: University of Rochester Press, 2003), pp. 207-8.
8. Rothstein *Public Health and the Risk Factor*, pp. 207-8.
9. Carl C. Seltzer and Seymour Jablon, "Army Rank and Subsequent Mortality by Cause: 23-Year Follow-up," *American Journal of Epidemiology* 105 (1977): 559-66.
10. Sidney Pell and C. Anthony D'Alonzo, "Acute Myocardial Infarction in a Large Industrial Population: Report of a 6-Year Study of 1,356 Cases," *JAMA* 185 (1963): 831-38.

Chapter 5

Explanations for the Emergence of the Coronary Heart Disease Pandemic

The emergence of coronary heart disease as a major cause of death at midcentury in the United States led to intense speculation as to its causes. Experts believed that it was a well known disease that had intensified in advanced societies because of changes in modern lifestyles. They used two types of changes to explain the causes of the increased number of cases. One involved broad changes in social conditions in these societies that affected personal lifestyles. The other involved changes in specific components of the lifestyles known as risk factors. Experts claimed that the disease would continue to produce illness and death in millions of persons annually until these conditions were modified. This analysis of both types of explanations found no chronological or other correspondence between the lifestyle changes and the emergence of this new form of coronary heart disease.

Any theory of causes of increases in mortality rates requires that the causal factors change before the increase in mortality rates. In the case of a pandemic, the causal factors must also change in the opposite direction before the decrease in mortality rates if they are true causal factors. In chronic diseases such as coronary heart disease, the time lags between the changes in the causal factors and the changes in mortality rates are typically multiple years. This chapter will describe changes in risk factors during the emerging phase of the pandemic and Chapter 9 will describe changes in risk factors during the declining phase of the pandemic.

Risk factors for pandemic coronary heart disease differed from those for normal coronary heart disease before the pandemic. This is clearly indicated by the substantial differences in the characteristics of the two forms of the disease. During the pandemic mortality rates were much higher in general, higher for men than women, and higher for older than younger age groups. Practically all research on coronary heart disease risk factors was undertaken during the pandemic, so that little research exists concerning risk factors for normal coronary heart disease before the pandemic.

An important consideration in evaluating coronary heart disease risk factors is that they have a cumulative effect. Persons with multiple risk factors have a much higher risk of disease than those with one or two risk factors. Changes in only one or two risk factors could not have produced the enormous increases in mortality rates that occurred during the coronary heart disease pandemic.

The basic broad changes in lifestyles that experts at midcentury believed caused the pandemic included greater consumption of heavily processed and allegedly unhealthy foods, decreased physical activity resulting from urbanization and new methods of personal transportation, and greater emotional stress caused by factors such as urban crowding and employment in large bureaucracies. Other broad risk factors were added as the pandemic progressed. These extremely general theories were not based on systematic research but rather on experts' beliefs about differences between societies with high and low coronary heart disease mortality rates.[1]

There are several reasons why broad changes in lifestyles could not have caused the increases in coronary heart disease rates. Broad lifestyle changes take decades to develop and diffuse throughout the population, but coronary heart disease mortality rates began to increase sharply in less than a decade. Changes in lifestyles tend to occur first among population groups with above average incomes that can afford to make the changes, but the highest coronary heart disease mortality rates were among the poor, not the wealthy.[2] Young persons adopt new lifestyles readily because they never learned the older lifestyles, while older persons tend to retain the lifestyles that they learned when they were young. If broad changes in lifestyles were responsible for the pandemic, younger persons should have experienced more rapid and greater relative increases in coronary heart disease mortality rates. However, older persons experienced the greatest relative increases in mortality rates and increases over longer periods of time.

The most frequently cited specific changes used to explain the emergence of the coronary heart disease pandemic involve individual behaviors and physiological conditions known as risk factors, which were described in Chapter 2. A set of important risk factors for coronary heart disease was specified in 1990 by the United States Department of Health and Human Services for its Healthy People program, which established objectives for the health of the American population in 2000. The objectives were later updated for 2010 and 2020. The specific risk factors considered important for coronary heart disease in the 2000 edition and subsequent revisions included: high blood cholesterol, dietary total fat intake, dietary saturated fat intake, high blood pressure, overweight, physical inactivity, and cigarette smoking. Diabetes was added as a risk factor in the revisions.[3]

Measuring the impact of specific risk factors on coronary heart disease during the height of the pandemic must consider the basic characteristics of all risk factors. The great majority of individuals with above average levels of particular risk factors have levels that are only slightly above average. They are only slightly more likely to develop the disease than persons with normal levels. Only a very small proportion of the population with above average levels of particular risk factors has levels that are high enough to produce a meaningful increase in the probability of developing the disease. In addition, practically all chronic disease risk factors have long latency periods and must be present in the individual for many years to produce disease.

For these reasons, a very large proportion of the total population had to be exposed to the postulated risk factors for long periods of time to produce the millions of additional cases of coronary heart disease that occurred annually during the peak of the pandemic. This makes it possible to measure changes in specific risk factors at the population level and relate them to changes in population coronary heart disease mortality rates.

The "diet-heart hypothesis" holds that patterns of food consumption are key risk factors for coronary heart disease, especially foods containing dietary cholesterol and saturated fats. Dietary cholesterol is present only in animal foods and saturated fats are present in animal foods and hydrogenated vegetable oils. Animal foods constitute a large part of the American diet and the most widely consumed animal foods include meats, chicken, eggs, dairy products, and fats such as butter. If the diet-heart hypothesis has merit, a very large proportion of American adults had to make major changes their diets by consuming more of these foods before and during the 1930s to produce the coronary heart disease pandemic.[4]

No meaningful changes occurred from the 1920s to the 1950s in the components of the American diet that could have produced large increases in coronary heart disease mortality rates. The standard measure of food intake is "foods available for consumption," which include consumed foods plus wastage and foods that were prepared but not consumed. Butter, lard, and shortening available for consumption per capita changed from 31 pounds in 1910 to 35 pounds in 1920, 40 pounds in 1930, 40 pounds in 1940, 34 pounds in 1950, and 29 pounds in 1957. Annual consumption of meats of all kinds changed from 146 pounds carcass weight per capita in 1910 to 136 pounds in 1920, 129 pounds in 1930, 142 in pounds 1940, 145 pounds in 1950, and 159 pounds in 1957. Eggs available for consumption changed from 306 per person in 1910 to 299 in 1920, 331 in 1930, 319 in 1940, 389 in 1950, and 360 in 1957. Chicken available for consumption changed from 16 pounds per person in 1910 to 14 pounds in 1920, 16 pounds in 1930, 14 pounds in 1940, 21 pounds in 1950, and 25 pounds in 1957. Consumption of dairy products changed little. Fluid milk and cream available for consumption remained unchanged at about 350 pounds per person per year from 1920 to 1957. Consumption of ice cream per person per year increased from 2 to 18 pounds and cheese increased from 4 to 8 pounds from 1920 to 1957.[5]

Many experts also claimed that specific nutritional components of foods, particularly dietary cholesterol and saturated fats, are important risk factors for coronary heart disease. Changes in the consumption of each of these items from 1909 to 1970 were too small to produce a substantial increase in coronary heart disease mortality rates. Dietary cholesterol per person per day available for consumption comprised 509 milligrams in 1909-13, 524 milligrams in 1925-29, 493 milligrams in 1935-39, 577 milligrams in 1947-49, 578 milligrams in 1957-59, and 556 milligrams in 1970. Saturated fats per person per day available for consumption were 50 grams in 1909-13, 53 grams in 1925-29, 53 grams in 1935-39, 54 grams in 1947-49, and 56 grams in 1970. Total animal and vegetable fats per person per day available for consumption were 125 grams in 1909-13, 135 grams in 1925-29, 133 grams in 1935-39, 131 grams in 1947-49, and 157 grams in 1970. The standard list of vitamins and minerals and the amount of protein available for consumption exhibited only small changes per person between 1909 and 1970.[6]

An increase in the number of calories consumed can produce obesity and diabetes, both risk factors for coronary heart disease. Total caloric intake per person per day changed little over the period, from 3490 calories in 1910 to 3290 calories in 1920, 3440 calories in 1930, 3350 calories in 1940, 3260 calories in 1950, 3140 calories in 1960, and 3350 calories in 1974.[7]

Thus the total diet of the American population did not change in ways that could have been responsible for the disease that became the single most important cause of death among middle-aged and older adults in the middle of the twentieth century. Most aspects of the diet did not change at all, others changed so as to produce small increases in risk, and still others changed to produce small decreases in risk. Many of the changes occurred after the emergence of the pandemic.

Experts have claimed that coronary heart disease rates are higher at any given time among persons with higher dietary total fat and saturated fat intake, regardless of their contribution to the rise of the pandemic. Many research studies have failed to support these hypothesized relationships. For example, a meta-analysis published in 2010 of 21 prospective cohort studies involving 347,000 persons measured the relationship between saturated fat intake and coronary heart disease. It concluded: "there is no significant evidence for concluding that dietary saturated fat is associated with an increased risk of" coronary heart disease or cardiovascular disease. Other reviews of the literature have produced similar findings.[8]

Many experts have also claimed that increased consumption of mass-produced prepared foods of low nutritional value served in restaurants, often called "fast foods," was partly responsible for the increase in coronary heart disease. This did not occur because fast food restaurants became popular in the 1970s, when it will be shown that coronary heart disease mortality rates were decreasing. In addition, the persons who consume fast foods most frequently are the young, while the greatest increases in coronary heart disease mortality rates occurred among the elderly. A 2005 population-based telephone survey of 4,311 adults in Michigan found that 37 percent of persons ages 18-24 ate at fast food restaurants at least two times a week compared to 18 percent of those ages 55-64.[9]

Many experts consider higher rates of cigarette smoking in the twentieth century to be a major cause of the increase in coronary heart disease mortality rates. Cigarette smoking has been proven to be a significant risk factor for coronary heart disease, several forms of cancer, and other diseases in prospective studies that followed thousands of smokers and nonsmokers in communities for many years. It has also been found that cigarette smokers experience a latency period of many years before the onset of serious disease. These studies are among the most methodologically sophisticated studies of risk factors ever undertaken. They have led to the acceptance of cigarette smoking as the single most important preventable cause of death and disease in all advanced countries.[10]

Even though cigarette smoking is a risk factor for coronary heart disease, the increase in smoking rates was not an important cause of the pandemic. To be a risk factor for the rise of the pandemic, smoking rates must increase before its emergence and decrease before its decline. This analysis will examine trends in cigarette smoking during the emergence of the pandemic and Chapter 9 will examine trends during the decline of the pandemic.

Widespread cigarette smoking in the United States occurred after the emergence of the coronary heart disease pandemic. Available data on cigarette smoking before midcentury consist of annual state revenues from cigarette taxes, which are converted to per capita sales. Cigarette consumption experienced its greatest increase during the 1940s, while the coronary heart disease pandemic began its rise about 1930 and was a serious national health problem by the 1940s. Per capita annual consumption of cigarettes increased from 477 in 1920 to 977 in 1930, 1349 in 1940, 2390 in 1950, 2645 in 1960, 2534 in 1970, 2752 in 1980, 2060 in 1990, 1551 in 2000, and 1001 in 2010.[11]

If cigarette smoking was a cause of the coronary heart disease pandemic, the rates of other chronic diseases affected by smoking should increase within a decade or so of the increase in coronary heart disease rates. The great majority of cases of lung cancer are caused by cigarette smoking in the United States and other advanced countries where smoking is widespread. However, lung cancer in white men and women became an important health problem several decades after the emergence of the coronary heart disease pandemic in the 1930s. Mortality rates per 1000 persons ages 55-64 from cancers of the respiratory system increased for white men from 0.5 in 1940 to 1.0 in 1950 and 1.5 in 1960 and for white women from 0.1 in 1940 to 0.2 in 1950 and 0.2 in 1960.[12]

A related factor that downgrades smoking as a risk factor in the emergence of the coronary heart disease pandemic is the younger ages of deaths of lung cancer victims than coronary heart disease victims. U.S. vital statistics indicate that one half or more those who died of lung cancer in 1968 were younger than age 65, including white and black men and women considered separately. By contrast, more than half of those who died of ischemic heart disease were 65 years of age or older, including white and black men and women considered separately.[13]

If smoking was an important risk factor for coronary heart disease, the younger ages of lung cancer deaths means that lung cancer mortality rates should have increased before the increase in coronary heart disease mortality rates. Instead, the increase in lung cancer mortality rates occurred long after the increase in coronary heart disease mortality rates.

Hypertension and diabetes have been proven to be important risk factors for coronary heart disease in individuals but they did not contribute to the rise of the coronary heart disease pandemic. The number of cases of coronary heart disease that occurred in persons with either of the two conditions was small. In 1955, of the 469,000 deaths that listed arteriosclerotic heart disease as the underlying cause, only 11 percent listed hypertensive disease as an associated cause and only 4 percent listed diabetes. Mortality rates from both hypertensive disease and its subcategory of hypertensive heart disease decreased steadily from 1949 to 1960 for white men, white women, nonwhite men, and nonwhite women in all age groups from 35-44 to 65-74. Mortality rates from diabetes decreased steadily from 1940 to 1960 for both white men and women in each age group from 35-44 to 65-74. Mortality rates from diabetes for nonwhite men and women of the same ages either decreased or remained the same from 1940 to the early 1950s and then increased to 1960.[14] Mortality rates from diabetes during this period were higher among women than men, but men had much higher coronary heart disease mortality rates than women.

Another fact that contradicts the causal role of most accepted risk factors in the pandemic is the large differences in coronary heart disease mortality rates among race and sex groups in each age group. If the hypothesized risk factors were major causes of the increased mortality rates, the prevalence of risk factors should vary in the population groups in proportion to the differences in their mortality rates. No evidence was found of such differences.

This analysis has demonstrated that changes in population risk factors cited in the Healthy People program did not produce the increase in coronary heart disease mortality rates that began in the 1930s. It will be shown subsequently that changes in these risk factors also did not produce the decreases in coronary heart disease mortality rates that occurred after 1970 in the United States.

Pandemics always emerge in specific geographic areas. It is therefore necessary to examine geographic differences in mortality rates during the emergence and peak of the pandemic. The widely separated and socially and economically diverse states of the United States are well suited to such an analysis.

References

1. For example, see Harvey Levenstein, *Fear of Food: A History of Why We Worry about What We Eat* (Chicago, IL: University of Chicago Press, 2012).
2. D.S. Grimes, "An Epidemic of Coronary Heart Disease," *Quarterly Journal of Medicine* 105 (2012): 511
3. National Center for Health Statistics, *Healthy People 2000 Review, 1998-99* (Hyattsville, MD: Public Health Service, 1999), pp. 148-49. For later versions, see http:www.healthypeople.gov.
4. Descriptions of the history of the diet-heart hypothesis can be found in Levenstein, *Fear of Food*, pp. 125-59; Ann F. La Berge, "How the Ideology of Low Fat Conquered America," *Journal of the History of Medicine and Allied Sciences* 63 (2008): 139-77; William G. Rothstein, *Public Health and the Risk Factor: A History of an Uneven Medical Revolution* (Rochester, NY: University of Rochester Press, 2003), pp. 295-342. For critical analyses of the diet-heart hypothesis, see George V. Mann, ed., *Coronary Heart Disease: The Dietary Sense and Nonsense* (London, Eng.: Janus, 1993); Uffe Ravnskov, *The Cholesterol Myths: Exposing the Fallacy that Saturated Fat and Cholesterol Cause Heart Disease* (Washington: NewTrends, 2000).
5. U.S. Bureau of the Census, *Statistical Abstract of the United States: 1958* (Washington, DC: 1958), p. 84.
6. Willis A. Gortner, "Nutrition in the United States, 1900 to 1974," *Cancer Research* 35 (1975): 3246-53. A rigorous study of patients in mental hospitals in Minnesota found no decrease in acute and silent myocardial infarctions and sudden deaths in 9423 patients who were fed a diet lower in saturated fats and cholesterol and higher in polyunsaturated fats compared to those fed a regular diet. The lack of difference in events occurred for all patients as well as those patients who were in the hospital for more than two years. The research methods were similar to those of a laboratory investigation The participants ate in common dining halls where they gave cards to the servers, who served them foods based on the two different diets. Serum cholesterol levels were measured periodically by blood tests and showed lower levels in those on the treatment diet. Cardiovascular events were measured by electrocardiograms after suspected events and autopsies where possible. This study is extremely rare for dietary intake studies in its ability to measure long-term consumption patterns directly and objectively rather than through participant recall. The study began in 1968 and lasted for more than four years. Ivan D. Frantz, Jr. et al, "Test of Effect of Lipid Lowering by Diet on Cardiovascular Risk," *Arteriosclerosis* 9 (1989): 129-35: Christopher E. Ramsden, Daisy Zamora, Sharon Majchrzak-Hong, Keturah R. Faurot, Steven K. Broste, Robert P. Frantz, John M. Davis, Amit Ringel, Chirayath M. Suchindran, Joseph R. Hibbeln, "Re-evaluation of the traditional diet-heart hypothesis: analysis of recovered data from Minnesota Coronary Experiment (1968-73)." BMJ. 2016 Apr 12;353:i1246. doi: 10.11.
7. Gortner, "Nutrition in the United States."

8. Patty W. Siri-Tarino, Qi Sun, Frank B. Hu, and Ronald M. Krauss, "Meta-Analysis of Prospective Cohort Studies Evaluating the Association of Saturated Fat with Cardiovascular Disease," *American Journal of Clinical Nutrition* 91 (2010): 535-46; U. Ravnskov, "The Questionable Role of Saturated and Polyunsaturated Fatty Acids in Cardiovascular Disease," *Journal of Clinical Epidemiology* 51 (1998): 443-60; Rothstein, *Public Health and the Risk Factor*, pp. 295-342.
9. Beth Anderson, et al, "Fast-food Consumption and Obesity among Michigan Adults," *Preventing Chronic Disease: Public Health Research, Practice, and Policy* 8 (2011) A71.
10. Richard Doll, Richard Petro, Jilliam Boreham, Isabelle Sutherland, "Mortality in Relation to Smoking: 50 Years' Observations on Male British Doctors," *BMJ* 328 (2004): 1519-28; Anthony J. Alberg and Jonathan M. Samet, "Epidemiology of Lung Cancer," *Chest* 123 (2003): suppl. 21S-49S.
11. *The Tax Burden on Tobacco: Historical Compilation* 46 (2011): 6 http://www.taxadmin.org/fta/tobacco/papers/Tax_Burden_2011.pdf (accessed March 14, 2015).
12. Robert D. Grove and Alice M. Hetzel, *Vital Statistics Rates in the United States, 1940-1960* (Washington, DC: National Center for Health Statistics, 1968), pp. 509-10.
13. Accessed at http://wonder/cdc.gov (accessed July 29, 2016).
14. Iwao M. Moriyama, Dean E. Krueger, and Jeremiah Stamler, *Cardiovascular Diseases in the United States* (Cambridge, MA: Harvard University Press, 1971), p. 90; Grove and Hetzel, *Vital Statistics Rates in the United States, 1940-1960*, pp. 466-74, 438-40.

Chapter 6

State Differences in Coronary Heart Disease Mortality Rates in the United States in 1950 and 1960

Major pandemics produce mortality rates that vary considerably among geographic regions, just as they do among population groups. About 1950 the coronary heart disease pandemic in the United States was more severe in states that had certain characteristics. As coronary heart disease mortality rates increased during the 1950s, the pandemic spread to the other states and made mortality rates more similar in all states. States with higher coronary heart disease mortality rates for men had higher rates for women, indicating the similarity of the pandemic in all states. The coronary heart disease mortality rates in the states had little correspondence to their mortality rates from all other causes. A striking geographic feature of the early years of the pandemic was the high mortality rates in a group of contiguous states in the northeastern region of the country.

The coronary heart disease pandemic, like all pandemics, was a social phenomenon that varied in severity among geographic regions. Comparisons of mortality rates among geographic regions are a valid scientific procedure because experts agreed that broad social changes played a key role in the pandemic and these vary by geographic regions. The 48 states of the United States are well suited to an analysis of factors associated with the pandemic in a very large geographic area in 1950 and 1960. The states varied in their coronary heart disease mortality rates and differed geographically, socially, culturally, and economically. They had a population of 135 million white persons in 1950 and provided statistical data at least the equal of any country in the world. The similarities of their health care systems reduce the possibility that differences in state mortality rates were caused by differences in diagnoses or reporting.

Comparisons of coronary heart disease mortality rates among states at mid-century are feasible only for white men and white women. The nonwhite population in many states was too small to produce accurate coronary heart disease mortality rates in those states. In 1950 nonwhite men, women, or both in 22 states had fewer than 20 age adjusted deaths due to arteriosclerotic heart disease. Furthermore, the nonwhite population was varied and included Negroes (the term used in the reports), Asians, American Indians, Eskimos, and others. Mexicans, Cubans, and Puerto Ricans were classified as white. Negroes comprised 96 percent of the total nonwhite population in the United States in 1950, but that proportion varied greatly by state. Negroes comprised less than half the nonwhite population in 13 of the 48 states in 1950.[1]

This analysis examines arteriosclerotic heart disease mortality rates and other state characteristics for white men and white women separately in the 48 states in the union in 1950 and the same states in 1960 using United States vital statistics described in Chapter 4. It considers each state to be an entity and disregards differences in the size of the populations and numbers of deaths in the states. Age adjusted mortality rates for each sex were used for analyses because they standardized the age distributions of the population of every state. Consequently variations in mortality rates among the states described in this analysis can not have been caused by differences in their age or sex distributions, two of the most important factors affecting coronary heart disease mortality rates.

It is essential to ensure that the same form of arteriosclerotic heart disease occurred in every state. One of the most distinctive characteristics of the pandemic was the much higher mortality rates of men than women. In both 1950 and 1960 men had considerably higher arteriosclerotic heart disease mortality rates than women in every state, demonstrating that the pandemic form of the disease occurred in all states.

Variations in arteriosclerotic heart disease mortality rates among the states are a significant issue. Practically every known pandemic has varied in intensity among geographic regions. In addition, most pandemics appear first in some geographic regions and then spread to others. Several factors have been found in many studies to affect state differences in mortality rates from a variety of diseases. One factor is the general health of the state population. States with higher mortality rates excluding arteriosclerotic heart disease may be more susceptible to new diseases such as pandemic coronary heart disease. Another factor is the per capita personal income of the residents of the state, because higher per capita incomes are typically associated with lower mortality rates. The proportion of the state population that lives in urban areas is important because urban-rural differences in overall mortality rates and mortality rates from specific diseases are common.

Considering first the general health of the population of the states, between 1950 and 1960 the states became healthier places to live as shown by the decreases in age adjusted total mortality rates excluding arteriosclerotic heart disease in every state. The mortality rates of the states became more similar, as shown by the decrease in the standard deviations. The average age adjusted white male total mortality rate excluding coronary heart disease per 1000 population in 1950 for all 48 states was 7.4, with a standard deviation of 0.53. In 1960 the average decreased to 6.6, with a standard deviation of 0.43. The standard deviation shows that in 1950 about two-thirds of the states had white male mortality rates that were no more than 0.53 deaths above or below the national average (i.e., rates between 6.9 and 7.9 deaths per 1000). In 1960 about two-thirds of the states had white male mortality rates that were no more than 0.43 deaths above or below the national average (i.e., between 6.2 and 7.0 deaths per 1000). For white women the average age adjusted mortality rate per 1000 population for all states in 1950 was 5.6, with a standard deviation of 0.41 deaths. In 1960 the average decreased to 4.6, with a much smaller standard deviation of 0.28 deaths.[2]

The arteriosclerotic heart disease pandemic continued to intensify and spread among the states from 1950 to 1960. Age adjusted mortality rates increased in 47 of the 48 states for white men and in all 48 states for white women. The pandemic became more severe in the states with originally lower rates, as shown by the increased similarity of the mortality rates of all the states as measured by the decrease in the standard deviations. In 1950 the average white male age-adjusted arteriosclerotic heart disease mortality rate per 1,000 population for all 48 states was 2.4, with a standard deviation of 0.41. In 1960 the average for all states for white men increased to 3.0, with a smaller standard deviation of 0.36. The average rate per 1000 for all states for white women in 1950 was 1.1, with a standard deviation of 0.27. For white women in 1960 the average increased to 1.4, with a standard deviation of 0.25.

The rise of the pandemic also increased the sex differences in the average arteriosclerotic heart disease mortality rates of the states. In 1950 the average state age adjusted arteriosclerotic heart disease mortality rate for white men per 1000 persons was 1.3 more deaths than for white women. In 1960 this increased to 1.6 more deaths. This provides further evidence that the sex difference was associated with the pandemic.

State differences in arteriosclerotic heart disease mortality rates provide useful information about the nature of the pandemic. The mortality rates of each state can be related to other characteristics of the states using Pearson correlation coefficients and the 48 states are adequate for these purposes. Pearson correlation coefficients measure the degree of correspondence between the arteriosclerotic heart disease mortality rate of each state and the numerical level of the other characteristic of the state being studied. The average correspondence for all states is then computed using a set of statistical techniques and is called a correlation coefficient. A positive correlation (states with higher arteriosclerotic heart disease mortality rates have higher scores on the other characteristic) produces a positive coefficient up to a maximum of +1.0, and a negative correlation (states with higher arteriosclerotic heart disease mortality rates have lower scores on the other characteristic) produces a negative coefficient up to a maximum of -1.0. The absence of any correlation between the two characteristics in the states produces a coefficient of 0.0.

Men had considerably higher arteriosclerotic heart disease mortality rates than women, and states with higher mortality rates for men also had higher rates for women. This indicates that the same factors that produced high mortality rates in men also produced high mortality rates in women. A very high degree of positive correspondence between the arteriosclerotic heart disease mortality rates of white men and white women occurred in the 48 states in both 1950 and 1960. The Pearson correlation coefficient between the age adjusted mortality rates of white men and white women in the 48 states in 1950 was $r = 0.92$ and that for 1960 was also $r = 0.92$. These extremely high correlations can be compared to the moderate correlations between the total mortality rates excluding arteriosclerotic heart disease of white men and white women. In 1950 these were $r = 0.52$ and in 1960 $r = 0.59$.

No consistent or meaningful relationship was found between the arteriosclerotic heart disease mortality rates of the individual states and their mortality rates from all other causes. Chronic diseases were the primary causes of adult deaths, with heart diseases, cancers, and stroke the three major causes.[3] For white men, the Pearson correlation coefficients for the correspondence between state age-adjusted arteriosclerotic heart disease rates and state age-adjusted total mortality rates excluding arteriosclerotic heart disease in the 48 states were very low and slightly negative: $r = -0.26$ in 1950 and $r = -0.20$ in 1960. The same coefficients for white women were very low and slightly positive: $r = 0.10$ in 1950 and $r = 0.24$ in 1960.

Two very common causes of geographic differences in mortality rates from many diseases are per capita personal income and urbanization, which are interrelated. Substantial variations were found among states in their per capita incomes. In 1950 the average state per capita personal income was $1401, the highest was $2132, and the lowest was $755; in 1960 the average state per capita personal income increased to $2057, the highest to $2856, and the lowest to $1204.[4]

States with higher per capital personal incomes were more likely to have higher age adjusted arteriosclerotic heart disease mortality rates. The correlations between the two were high in 1950 and decreased somewhat in 1960 because of the spread of the pandemic to lower income states. This suggests that the factors that produced the pandemic occurred first in higher income states but were not limited to those states. The Pearson correlation coefficients for the relationship between white male age-adjusted arteriosclerotic heart disease mortality rates and per capita personal income in the 48 states were $r = 0.65$ in 1950 and $r = 0.40$ in 1960. The coefficients for white women were very similar to those of the men: $r = 0.61$ in 1950 and $r = 0.48$ in 1960.

A weak relationship was found between the per capita personal incomes of the states and their age adjusted total mortality rates excluding arteriosclerotic heart disease for each sex considered separately. In 1950 there was no relationship between the two, with Pearson coefficients of $r = -0.03$ for men and $r = -0.09$ for women. In 1960, lower income states were very slightly more likely to have higher mortality rates, with $r = -0.24$ for white men and $r = -0.24$ for white women.

The 1940 finding of higher coronary heart disease mortality rates in towns and cities than rural areas continued in 1950 but diminished in 1960 as the pandemic spread throughout the country. A positive relationship was found in 1950 between a state's age adjusted arteriosclerotic heart disease mortality rate and the percentage of the state's population that lived in urban areas. Urban was defined as places of at least 2500 inhabitants plus densely settled fringe areas around cities of at least 50,000 population. These varied among states in 1950 from low of 27 percent to a high of 88 percent. In 1950 the Pearson correlation coefficient for white men was a high $r = 0.72$ and for white women it was a similar $r = 0.68$. In 1960 the percentage of the population living in urban areas varied among states from a low of 35 percent to a high of 89 percent. The Pearson coefficient for the correspondence between the age-adjusted arteriosclerotic mortality rates and the urban percentage of the population of the states for white men was $r = 0.58$ and for white women was $r = 0.39$.[5]

Another issue in a pandemic is the stability of state rankings of arteriosclerotic heart disease mortality rates over time. If changes in the mortality rates of each state during the pandemic were due to basic causal factors operating at the national level, the rankings of the mortality rates of the states should be stable over periods of time. If state level factors were more important in producing changes in mortality rates, the rankings should fluctuate more over time because of differences among the states. Basic causal factors at the national level were responsible for changes in arteriosclerotic heart disease mortality rates as shown by the very stable rankings of the mortality rates between 1950 and 1960. The Pearson correlation coefficient between the 1950 and 1960 age adjusted arteriosclerotic heart disease mortality rates of the states for white men was a very high $r = 0.82$, as was that for white women, $r = 0.88$.

Stable state rankings also occurred for total mortality rates from all other causes. This is more likely due to long-term factors operating in the states, as will be shown in Chapter 10. Considering the correspondence between the 1950 and the 1960 total mortality rates excluding arteriosclerotic heart disease of the states, the Pearson correlation coefficient for white men was $r = 0.83$ and that for white women was $r = 0.73$.

This analysis of state mortality rates has found that higher average arteriosclerotic heart disease mortality rates occurred in states with higher incomes and in states with higher proportions of urban residents, two state characteristics that are interrelated. The studies of individual mortality rates reported in Chapter 5 were conducted primarily in urban areas and found that persons in lower socioeconomic groups had higher mortality rates than persons in higher groups. This suggests that lower socioeconomic groups in urban areas were the most frequent victims of the disease. Some studies in urban areas early in the pandemic found that to be the case.[6]

States with the Highest and Lowest Arteriosclerotic Heart Disease Mortality Rates in 1950 and 1960

Pandemics have typically produced their highest mortality rates in specific geographic locations of the affected countries. This analysis identified a group of contiguous states in the northeastern United States that had the highest mortality rates for men and women in both 1950 and 1960. Other characteristics of those states varied considerably, which suggests that the coronary heart disease pandemic had a geographic component that was independent of other factors.

In order to identify the states with the highest mortality rates when the coronary heart disease pandemic was reaching its peak in 1950 and 1960, lists of mortality rates by state were constructed to obtain the ten states with the highest age adjusted arteriosclerotic heart disease mortality rates for white men and the ten states with the highest rates for white women. Seven states were on both the male and female lists in 1950 and eight states were on both lists in 1960. These were considered the states that were most severely affected by the peak of the pandemic.

In 1950 the seven states with very high arteriosclerotic heart disease mortality rates for men and women were clustered in a small geographic region with every state adjoining one or more states on the list (Table 6.1). Their locations strongly indicate that the geographic component of the pandemic operated somewhat independently of other factors. All seven states were in the northeastern United States, including four New England states (Massachusetts, Connecticut, Rhode Island, and New Hampshire) and three mid-Atlantic states (New York, New Jersey, and Delaware). The seven contiguous states had a combined population of 28 million in a total United States population of 151 million, or 19 percent of the total population. The white male arteriosclerotic heart disease rates per 1000 in the seven highest states varied from 3.5 to 2.9 compared to the average of 2.4 for all states. The white female arteriosclerotic heart disease mortality rates per 1000 in the seven highest states varied from 1.9 to 1.4 compared to the average of 1.1 for all states.[7]

Additional evidence of a specific geographic component of the pandemic in 1950 is provided by the variations in other characteristics of the states with the highest arteriosclerotic heart disease mortality rates (Table 6.1). The states were close to the national average for mortality rates from other causes for both men and women. The per capita incomes of all except one state were above the national average, but the incomes varied considerably among the states. All of the states had an above average proportion of the population residing in urban areas, but considerable variations existed among the states.

Table 6.1 Characteristics of States with Highest Arteriosclerotic Heart Disease Mortality Rates for White Men and Women, 1950 and 1960 (State age adjusted mortality rates per 1,000 persons)

State	*Arteriosclerotic Heart Disease Mortality*		*Other Causes Mortality*		*State per Capita Income ($)*	*State Percent Urban (%)*
	White Men	*White Women*	*White Men*	*White Women*		
1950						
New York	3.5	1.9	6.8	5.4	1873	85
Rhode Island	3.2	1.6	7.2	5.9	1605	85
New Hampshire	3.1	1.5	7.1	5.7	1323	58
Connecticut	3.0	1.6	6.4	5.1	1875	77
Massachusetts	3.0	1.6	6.7	5.3	1633	84
Delaware	3.0	1.4	8.0	6.3	2132	62
New Jersey	2.9	1.6	7.2	5.8	1834	86
all state average	2.4	1.1	7.4	5.6	1401	55
1960						
Rhode Island	3.8	2.0	5.9	4.4	2219	86
New York	3.7	2.0	6.0	4.5	2748	85
Massachusetts	3.6	1.8	6.5	4.7	2461	84
Illinois	3.6	1.8	6.3	4.5	2649	81
New Hampshire	3.5	1.7	6.6	4.7	2144	58
Pennsylvania	3.5	1.8	6.6	4.9	2241	72
New Jersey	3.5	1.8	6.0	4.7	2707	89
Maine	3.5	1.7	6.8	5.0	1844	51
all state average	3.0	1.4	6.6	4.6	2057	62

Sources:

Mortality: Robert D. Grove and Alice M. Hetzel, *Vital Statistics Rates in the United States, 1940-1960* (Washingon, DC: National Center for Health Statistics, 1968), pp. 663-66, 720-22

Per capital income and percent urban: U.S. Bureau of the Census, *Statistical Abstract of the United States, 1970* (Washington, DC: 1970), pp. 17, 320-21.

The 1960 group of the eight states with very high arteriosclerotic heart disease mortality rates included five of the 1950 states (New York, Rhode Island, Massachusetts, New Hampshire, and New Jersey) (Table 6.1). The tight clustering continued because two of the three new states (Pennsylvania and Maine) adjoined one or more of the five states that were on both the 1950 and 1960 lists. Both new states had high coronary heart disease mortality rates in 1950. One new state, Illinois, was not geographically contiguous to the other states. The eight states on the 1960 list had a combined population of 52 million in a total United States population of 180 million, or 29 percent. The two states not on the 1960 list that had been on the 1950 list (Connecticut and Delaware) continued to have very high coronary heart disease mortality rates. The average white male arteriosclerotic heart disease mortality rate per 1000 in the eight states with the highest rates varied from 3.8 to 3.5 compared to 3.0 for all states. The average white female rate in the states with the highest rates varied from 2.0 to 1.7 compared to 1.4 for all states.[8]

The other characteristics of the states with the highest arteriosclerotic heart disease rates in 1960 also support the existence a specific geographic component of the pandemic. These characteristics varied considerably, as in the 1950 list. All had mortality rates from all other causes that were close to the average for all states. All except one state had above average per capita incomes, but the incomes varied considerably among the states. Six of the eight states had an above average proportion of their populations that were urban.

A similar process was used to select the states with the lowest arteriosclerotic heart disease mortality rates in 1950 and 1960 (Table 6.2). Lists were constructed in each year of the ten states with the lowest age adjusted white male arteriosclerotic heart disease mortality rates and the ten states with the lowest white female rates. Eight states were included on both lists in 1950 and nine states were included on both lists in 1960.

The states with very low arteriosclerotic heart disease mortality rates were much more geographically dispersed in both 1950 and 1960 than the states with the highest rates. In 1950 the states with very low rates were located in three widely separated geographic areas. Four adjoining southern states were near the Mississippi River (Arkansas, Tennessee, Alabama, and Oklahoma). Two adjoining states were in the southwest (New Mexico and Arizona) and two adjoining states (Nebraska and Wyoming) were in the north central region. The age adjusted arteriosclerotic heart disease mortality rates per 1000 white men in the states with the lowest rates varied from 1.7 to 2.1 compared to an average of 2.4 for all states. The same rates for women varied from 0.7 to 0.9 in the states with the lowest rates compared to an average of 1.1 for all states.

The states with very low arteriosclerotic heart disease mortality rates varied considerably in their other characteristics, as occurred in the states with very high rates. The mortality rates from all other causes of the states were similar to the national average. The per capita incomes of the states tended to be below average but varied considerable. Most of the states had below average percentages of urban residents but the proportions varied considerably.

In 1960, the states with very low arteriosclerotic heart disease mortality rates continued to be located over a broad geographic area. Five of the nine states continued from the 1950 list (New Mexico, Alabama, Oklahoma,

Wyoming, and Arizona) and each of the four new states adjoined one of the five 1950 states. The states on the 1960 list included two adjoining southern states east of the Mississippi River (Alabama and Mississippi), two adjoining southern states west of the Mississippi River (Oklahoma and Texas), and five adjoining states in or near the Rocky Mountains region (Idaho, Wyoming, Utah, Arizona, and New Mexico). The arteriosclerotic heart disease mortality rates per 1000 white men in the nine states with the lowest rates varied from 2.0 to 2.7 compared to an average of 3.1 for all states. The same rates per 1000 white women varied from 1.0 to 1.2 compared to an average of 1.4 for all states.

Table 6.2 Characteristics of States with Lowest Arteriosclerotic Heart Disease Mortality Rates for White Men and Women, 1950 and 1960 (State age adjusted mortality rates per 1,000 persons)

State	*Arteriosclerotic Heart Disease Mortality*		*Other Causes Mortality*		*State Per Capita Income* ($)	*State Percent Urban* (%)
	White Men	*White Women*	*White Men*	*White Women*		
1950						
New Mexico	1.7	0.7	8.5	6.9	1177	50
Arkansas	1.8	0.8	8.8	4.9	825	33
Tennessee	1.9	0.8	7.8	5.9	994	44
Nebraska	2.0	0.8	7.0	5.4	1490	47
Alabama	2.0	0.9	7.6	5.5	880	44
Oklahoma	2.0	0.8	7.0	5.0	1143	51
Wyoming	2.0	0.8	8.0	5.6	1668	50
Arizona	2.1	0.7	8.8	5.5	1330	61
all state average	2.4	1.1	7.4	5.6	1401	56
1960						
New Mexico	2.0	1.0	7.3	5.2	1888	66
Utah	2.4	1.0	6.3	4.4	1968	75
Arizona	2.6	1.2	7.0	4.2	2032	75
Mississippi	2.6	1.2	6.9	4.5	1204	38
Wyoming	2.7	1.1	7.0	5.0	2261	57
Texas	2.7	1.1	6.5	4.4	1924	75
Alabama	2.7	1.2	7.0	4.5	1488	55
Idaho	2.7	1.0	6.2	4.3	1850	48
Oklahoma	2.7	1.1	6.7	4.4	1801	63
all state average	3.1	1.4	6.6	4.6	2057	62

Sources: See Table 6.1

Other characteristics of the states with the lowest rates in 1960 were not closely related to their arteriosclerotic heart disease mortality rates, as in 1950. Their mortality rates for all other causes continued to be about average. Their per capita incomes were usually below the national average but varied considerably. Their percentages of urban residents varied from above to below average.

This analysis of arteriosclerotic heart disease mortality rates in the states of the United States in 1950 and 1960 demonstrated that the pandemic intensified and spread to more states over the decade. The mortality rates of the individual states were not related to their mortality rates from all other causes. A very high degree of correspondence existed between the mortality rates of men and women in each state, demonstrating the existence of the same causal factors in all states. States that were urban and had higher per capita incomes tended to have higher arteriosclerotic heart disease mortality rates. A geographic region with high arteriosclerotic heart disease mortality rates was located in a contiguous group of states in the northeastern section of the country. No comparable contiguous region was found consisting of the states with the lowest arteriosclerotic heart disease mortality rates. This suggests that the causal factors of the pandemic were more severe in particular locations but that no specific regions had low levels of the causal factors.

The pandemic developed about the same time in other many countries. Canada and England and Wales are particularly appropriate because their high standards of living made them similar in many respects to the United States.

References

1. Robert D. Grove and Alice M. Hetzel, *Vital Statistics Rates in the United States, 1940-1960* (Washington, DC: National Center for Health Statistics, 1968), pp. 43, 49, 720-2; U.S. Bureau of the Census, *Statistical Abstract of the United States: 1958* (Washington, DC: 1958), pp. 30-31.
2. The state mortality rates for 1950 and 1960 throughout this chapter are provided in or calculated using data from Grove and Hetzel, *Vital Statistics Rates in the United States, 1940-1960*, pp. 663-66, 720-22.
3. Grove and Hetzel, *Vital Statistics Rates in the United States, 1940-1960*, p. 79.
4. State per capita income for 1950 and 1960 was obtained from U.S. Bureau of the Census, *Statistical Abstract of the United States: 1970*. Washington, DC: 1970), p. 320.
5. U.S. Bureau of the Census, *Statistical Abstract of the United States: 1958*, pp. 2, 23; U.S. Bureau of the Census, *Statistical Abstract of the United States: 1970*. p. 17.
6. William G. Rothstein, *Public Health and the Risk Factor: A History of an Uneven Medical Revolution* (Rochester, NY: University of Rochester Press, 2003), pp. 206-7.
7. U.S. Bureau of the Census, *Statistical Abstract of the United States: 1958*, p. 10.
8. U.S. Bureau of the Census, *Statistical Abstract of the United States: 1970*. p. 12.

Chapter 7

The Emergence of the Coronary Heart Disease Pandemic in Canada and England and Wales

Canada and England and Wales have available vital statistics in English that permit detailed descriptions of the emergence and peak of the coronary heart disease pandemic in the two countries. The similar characteristics of the pandemic in both countries and the United States indicate the operation of the same causal factors in all three countries. Coronary heart disease mortality rates began to increase about the same time in all three countries. Men had higher mortality rates than women and older age groups higher mortality rates than younger age groups in the three countries. The provinces of Canada varied among themselves in their mortality rates in ways similar to the states of the United States. Mortality rates in Canada at the peak of the pandemic were slightly lower than in the United States and those in England and Wales were considerably lower.

This analysis of the emergence of the pandemic outside the United States is limited to Canada and England and Wales because appropriate vital statistics in the English language for many other advanced countries became available later in the century. Those countries will be analyzed in detail in Chapter 11 to document the peak and decline of the coronary heart disease pandemic.

Coronary Heart Disease Mortality Rates in Canada

The history of coronary heart disease in Canada from about 1920 to about 1960 was very similar to that in the United States and demonstrates the operation of the same causal factors in both countries. The populations were becoming healthier in both countries. Both countries experienced increases in coronary heart disease mortality rates in all adult age groups and much higher mortality rates for men than women and for older than younger age groups. The proportion of all deaths that were caused by coronary heart disease rose steadily in both countries and reached very high levels by midcentury. Canadian provinces exhibited differences in their mortality rates that were similar in many respects to those of the states in the United States.

Canada is an advanced country with a large population and a political structure that is very similar to the United States: a federal government and provinces with considerable political autonomy, geographic dispersion, and social, economic, and cultural diversity. In 1958 Canada had about 17 million inhabitants and a per capita gross national product of $1979, second only to the $2602 of the United States, with no other country having a per capita gross national product greater than $1520. During the 1950s slightly more than 70 percent of the total Canadian population lived in Ontario, Quebec, and British Columbia. Published vital statistics for Canada and its provinces before midcentury are rudimentary.[1]

Canadian coronary heart disease mortality rates from 1921 to 1949 are most useful for the province of Ontario. Ontario had 2.9 million residents in 1921 and 4.6 million in 1951, about one-third the total population of Canada in both years. Changes in the crude mortality rates of the province over time were not standardized for changes in the age distribution of its population. However, the increases in coronary heart disease mortality rates were so great that it is highly unlikely that they could have been caused by a larger proportion of middle-aged and older persons in the population.[2]

Crude coronary heart disease mortality rates in Ontario increased slightly from 1921 to 1930 and by very large amounts from the 1930s to 1949. Men experienced greater increases in mortality rates than women. For men, deaths due to "angina pectoris" increased slightly from 0.8 percent of 18,062 total deaths in all ages in 1921 to 2.6 percent of the 19,827 deaths in 1930. For women, the percentages increased slightly from 0.4 percent of 16,489 deaths in 1921 to 1.2 percent of 17,486 deaths in 1930. From 1935 to 1949 coronary heart disease deaths were listed in the new and expanded category of "angina pectoris and diseases of the coronary arteries." This period witnessed an extraordinary increase in mortality rates from the disease. In 1935 the new category was responsible for 9.5 percent of the 19,281 total male deaths and 5.6 percent of the 17,036 total female deaths in the province. In 1940 coronary heart disease produced 15.1 percent of the 20,923 total male deaths and 8.6 percent of the 17,580 total female deaths, and in 1949 17.0 percent of the 24,123 total male deaths and 9.1 percent of the 19,256 total female deaths.[3]

These extraordinary increases in coronary heart disease mortality rates in Ontario are supported by a study of samples of between 2,500 and 5,000 death certificates of men ages 45-64 in Ontario in each census year from 1901 to 1961. In order to deal with the problem of misdiagnosis, the authors used three definitions of coronary heart disease that varied in inclusiveness.

No clear evidence of an increase in coronary heart disease mortality rates was found from 1901 to 1931, but mortality rates increased by two or more times from 1931 to 1961 for each of the three definitions, with the greatest increases for the most narrow definition.[4]

Vital statistics by age and sex are available for the entire Canadian population beginning at midcentury. From 1951 to 1961, the Canadian adult population became steadily healthier (Table 7.1). For example, total mortality rates excluding arteriosclerotic heart disease per 1000 persons ages 45-49 decreased for men from 4.4 to 3.5 and for women from 4.0 to 2.8. For those ages 65-69 the mortality rates per 1000 for men decreased from 22.4 to 20.7 and those for women from 18.0 to 14.1.

Table 7.1 Canada Arteriosclerotic Heart Disease and Total Mortality Rates by Age and Sex, 1951 and 1961
(Rates per 1,000 persons)

Age group	*Arteriosclerotic Heart Disease Mortality*		*Total Mortality*	
	Male	*Female*	*Male*	*Female*
1951				
45-49	1.9	0.5	6.3	4.5
50-54	3.6	0.9	10.3	6.4
55-59	5.9	2.1	16.2	10.2
60-64	9.2	4.1	24.4	16.1
65-69	12.6	6.9	35.1	24.9
1961				
45-49	2.3	0.4	5.8	3.2
50-54	4.0	0.9	9.6	5.3
55-59	6.5	1.8	15.1	8.0
60-64	10.1	3.7	24.0	12.8
65-69	14.9	7.4	35.7	21.4

Sources: Number of arteriosclerotic heart disease and total deaths:
1951: Canada Dominion Bureau of Statistics, *Vital Statistics Statistisque de L'Etat Civil 1951* (Ottawa: Cloutier, 1954) pp. 261, 265
1961: Canada Dominion Bureau of Statistics, *Vital Statistics Statistisque de L'Etat Civil 1961* (Ottawa: Duhamel, 1963), pp. 152-53, 158-59.
Population:
1951: Canada Dominion Bureau of Statistics, *Canada Year Book 1952-53* (Ottowa, Canada: Cloutier, 1953), p. 145.
1961: Canada Dominion Bureau of Statistics, *Canada Year Book 1962* (n.p.: n.d.), p. 1202. http:www65.statcan.gc.ca (Accessed Feb. 27, 2015)

Mortality rates in Canada from "arteriosclerotic and degenerative heart diseases" from 1951 to 1961 demonstrated patterns similar to those in the United States. Mortality rates for men increased for all age groups, with older men having higher rates and larger increases (Table 7.1). For example, the male mortality rates per 1000 at ages 45-49 increased from 1.9 to 2.3 and at ages 65-69 from 12.6 to 14.9. Mortality rates for women were lower and largely stable from 1951 to 1961 with very low rates below age 60 and higher rates at older ages. For example, female arteriosclerotic heart disease mortality rates per 1000 at ages 45-49 were 0.5 in 1951 and 0.4 in 1961 and at ages 65-69 they were 6.9 in 1951 and 7.4 in 1961. The differences in arteriosclerotic heart disease mortality rates between older and younger age groups and between men and women increased somewhat between 1951 and 1961.

The proportion of total mortality in Canada that was caused by arteriosclerotic heart disease increased for both men and women between 1951 and 1961, partly due to the decrease in mortality rates from other causes. The proportions of all male deaths that were caused by arteriosclerotic heart disease at ages 45-49 increased from 30 to 39 percent, at ages 55-59 increased from 37 percent to 43 percent, and at ages 65-69 from 36 to 42 percent. The proportions of all female deaths that were caused by arteriosclerotic heart disease at ages 45-49 increased from 11 percent in 1951 to 12 percent in 1961, at ages 55-59 from 20 percent in 1951 to 22 percent in 1961, and at ages 65-69 from 28 percent in 1951 to 34 percent in 1961.

Thus the Canadian experience from the 1920s to 1960 was very similar to that of the United States and demonstrated the operation of the same causal factors in both countries. Coronary heart disease mortality rates increased after 1930 for all age groups and increased more for men than women. The rates were relatively high for men at all ages while women had low rates in the middle ages and higher rates at older ages. It will be shown in Chapter 11 that the mortality rates in Canada during the pandemic were slightly lower than those in the United States. Another difference between the two countries was that arteriosclerotic heart disease mortality rates for older persons increased less from 1950 to 1960 in Canada than in the United States. Mortality rates for younger persons did not increase from 1950 to 1960 in both countries.

The Canadian provinces varied substantially in their crude arteriosclerotic heart disease mortality rates, indicating that geographic factors were as important and independent of other factors in Canada as in the United States. In 1952 the average crude arteriosclerotic heart disease rate per 1000 population for all provinces was 2.8 for men and 1.7 for women in a total Canadian population of 7.3 million men and 7.1 million women. Of the 8 provinces with at least 200,000 inhabitants, three were atypical. Two provinces had exceptionally high mortality rates: Ontario with rates of 3.6 for men and 2.3 for women in a population of 2.5 million men and 2.4 million women, and British Columbia with rates of 3.7 and 2.1 in a population of 613,000 men and 585,000 women. Quebec had the lowest crude rates per 1000 population: 2.1 for men and 1.2 for women in a population of 2.1 million men and 2.1 million women. These substantial differences in crude arteriosclerotic heart disease mortality rates in provinces with large populations are very unlikely to be due to differences in the age distributions of their populations.[5]

The two provinces with the highest arteriosclerotic heart disease mortality rates (Ontario and British Columbia) and the province with the lowest rates (Quebec) had total crude mortality rates excluding arteriosclerotic heart disease in 1952 that were unrelated to their arteriosclerotic heart disease mortality rates. The rates from all other causes per 1000 population for all of Canada were 7.1 for men and 5.8 for women. Those for Ontario were 6.9 for men and 5.9 for women, those for British Columbia were 8.4 for men and 5.9 for women, and those for Quebec were 7.4 for men and 6.1 for women.[6]

The provinces with the highest and lowest crude arteriosclerotic heart disease mortality rates remained unchanged in 1960. The average crude arteriosclerotic heart disease mortality rates per 1000 population for all provinces were 3.0 for men and 1.8 for women. The highest rates in the large provinces continued to be for Ontario, with rates of 3.6 for men and 2.3 for women, and for British Columbia, with rates of 3.9 for men and 2.2 for women. Quebec again had the lowest rates, which were 2.3 for men and 1.3 for women. These rates continued to be unrelated to the crude total mortality rates excluding arteriosclerotic heart disease of the provinces in 1960. The total mortality rates excluding arteriosclerotic heart disease per 1000 for Canada were 6.1 for men and 4.8 for women. The rates for men in the three provinces were 6.0 in Ontario, 7.2 in British Columbia, and 5.6 in Quebec. The rates for women were 5.0 in Ontario, 5.0 in British Columbia, and 4.6 in Quebec.[7]

The per capita incomes of the three provinces were related to some extent to their arteriosclerotic heart disease mortality rates. In 1960 Ontario and British Columbia, the two provinces with the highest crude arteriosclerotic heart disease mortality rates, had the highest per capita incomes in Canada. Quebec, the province with the lowest mortality rate, had a per capita income that was slightly below average. The average per capita income in Canada in 1961 for men was $3999 and for women was $1651. In that year Ontario and British Columbia had the highest per capita incomes for men of all provinces, $4355 and $4177 respectively. They had per capita incomes for women that were close to the national average: $1747 and $1652 respectively. Quebec's per capita income for men of $3870 was slightly below the national average but much above that of the lowest income provinces, which had average per capita incomes of less than $3200. Its average per capita income for women of $1703 was also above average.[8]

The geographic patterns of arteriosclerotic heart disease mortality rates in the provinces of Canada thus had similarities and differences with the states of the United States. A high degree of stability occurred between the 1950 and 1960 rankings of the provinces and states respectively. The highest arteriosclerotic heart disease mortality rates in Canada occurred in the two provinces with the highest per capita incomes for men. In the United States the states with higher arteriosclerotic heart disease mortality rates were very likely to have above average per capita incomes. The provinces in Canada and the states in the United States with the lowest mortality rates did not have the lowest per capita incomes. A difference between the countries was that the American states with the highest arteriosclerotic heart disease mortality rates in 1950 and 1960 were located in a contiguous geographic area (with one exception in 1960) while the two Canadian provinces with the highest mortality rates were located two thousand miles apart.

Coronary Heart Disease Mortality Rates in England and Wales

England and Wales is another advanced country that experienced the coronary heart disease pandemic. It also has a detailed and impressive published description of vital statistics during the period. Its experiences during the emergence and peak of the pandemic were similar in many respects to the United States and Canada, which demonstrates the presence of the same causal factors in all three countries. The population of England and Wales was becoming steadily healthier. The timing of the increase in its mortality rates was similar to the United States and Canada. Men and older age groups had higher mortality rates than women and younger age groups in all three countries. The higher mortality rates of men than women also supported the accuracy of mortality reporting.

The per capita gross national product of the United Kingdom (including Scotland) in 1958 was $1254, which ranked seventh in the world. The mortality statistics to be analyzed here are for England and Wales, excluding Scotland. In 1931 the population of England and Wales was 40.0 million, which increased to 46.2 million in 1961.[9]

During the emergence and peak of the coronary heart disease pandemic from 1931-35 to 1961-65, all age groups of the population of England and Wales became healthier, with greater decreases in total mortality rates among the elderly (Table 7.2). For example, annualized average total mortality rates excluding coronary heart disease per 1000 for men ages 35-44 decreased from 5.2 in 1931-35 to 2.0 in 1961-65 and for ages 65-74 from 45.1 to 38.3. The rates per 1000 for women ages 35-44 decreased from 4.1 to 1.7 and for ages 65-74 from 34.1 to 23.4.

Coronary heart disease mortality rates increased by small amounts in the 1930s (Table 7.2). Male deaths from "diseases of the coronary arteries and angina pectoris" per 1000 population at ages 45-54 increased from 1.0 in 1931 to 1.3 in 1941 and at ages 65-74 from 11.6 to 13.9. Mortality rates for women at ages 45-54 remained stable over the period at 0.6 and at ages 65-74 increased from 8.8 in to 9.7.

Table 7.2 England and Wales Coronary Heart Disease and Total Mortality Rates by Age and Sex, 1931-1961
(Rates per 1,000 persons)

	Male				*Female*			
	35-44	*45-54*	*55-64*	*65-74*	*35-44*	*45-54*	*55-64*	*65-74*
Diseases of the coronary arteries and angina pectoris								
1931	0.2	1.0	3.3	11.6	0.2	0.6	2.3	8.7
1941	0.3	1.3	4.5	13.9	0.1	0.6	2.3	9.7
Arteriosclerotic heart disease including coronary disease								
1951	0.3	1.7	5.6	16.5	0.1	0.4	1.8	7.9
1961	0.5	2.1	6.4	15.7	0.1	0.4	1.7	6.4
Five-year annualized average total mortality rates								
1931-35	5.4	11.2	23.6	56.7	4.3	8.0	17.0	42.8
1941-45	4.8	9.9	23.1	51.7	3.3	6.4	14.0	36.0
1951-55	2.7	7.9	22.5	54.6	2.1	4.9	11.8	33.1
1961-65	2.5	7.4	21.7	54.0	1.8	4.4	10.6	29.8

Sources: John Charlton, "Trends in All-Cause Mortality: 1841-1994," p. I:23 and John Charlton, et al, "Cardiovascular Diseases," p. II:65, in *The Health of Adult Britain, 1841-1994*, ed. John Charlton and Mike Murphy, 2 vols. (London: Office of National Statistics, 1997).

The 1940s witnessed increases in the mortality rates of men but not of women, using the 1930s category in 1941 and the new category of arteriosclerotic heart disease in 1951. At ages 45-54 mortality rates per 1000 for men increased from 1.3 in 1941 to 1.7 in 1951 while those for women decreased from 0.6 to 0.4. At ages 65-74 the rates for men increased from 13.9 in 1941 to 16.5 in 1951 while those for women decreased from 9.7 to 7.9.

The period from 1951 to 1961 was one of general stability in arteriosclerotic heart disease mortality rates for both sexes. For example, male arteriosclerotic heart disease mortality rates per 1000 for ages 45-54 increased from 1.7 in 1951 to 2.1 in 1961, and for ages 65-74 decreased from 16.5 in 1951 to 15.7 in 1961. Those for women ages 45-54 were 0.4 in 1951 and 1961, and for ages 65-74 the rates decreased from 7.9 in 1951 to 6.4 in 1961.

Throughout the three decade period, men had higher mortality rates than women in every age group. The rates for women were quite low at younger ages but increased considerably at older ages. Both patterns were similar to those in the United States and Canada.

The proportion of all deaths that were caused by coronary heart disease in England and Wales increased generally from 1931 to 1961, with very large increases for men and mixed changes for women. At ages 45-54, the proportions for men increased from about 9 percent in 1931 to about 28 percent in 1961. For women of the same ages the proportions increased from about 8 percent in 1931 to 9 percent in 1961. At ages 65-74 the proportions for men increased from about 20 in 1931 to about 29 percent in

1961 and those for women from about 20 percent in 1931 to about 24 percent in 1951 and then decreased to about 21 percent in 1961. The increases were due in part to the decreases in mortality rates from all other causes.

Urban areas in England and Wales in 1950-52 had higher mortality rates than rural areas and a greater difference in the mortality rates of men and women. This pattern was similar to their total mortality rates.[10]

Conclusion

The many similarities in the trends and patterns of coronary heart disease mortality rates in the United States, Canada, and England and Wales from about 1930 to about 1960 indicate that the same causal factors operated in all three countries. The countries were becoming healthier as shown by their steadily decreasing total mortality rates excluding coronary heart disease. Coronary heart disease mortality rates were higher for men than women in the three countries, which also indicated the accuracy of the diagnoses. All age groups of men in the three countries and of women in the United States and Canada experienced increases in their coronary heart disease mortality rates for most of the period. The rates of older men and women increased more than those of younger men and women. All three countries experienced their peak mortality rates for younger age groups beginning in the 1950s. Geographic differences in coronary heart disease mortality rates occurred in the three countries.

A major difference among the three countries was the differences in the levels of their mortality rates during the peak of the pandemic at midcentury: the United States had the highest mortality rates, Canada slightly lower mortality rates, and England and Wales the lowest rates of the three countries. In addition, mortality rates for older age groups in the United States continued to increase in the 1960s while those in Canada and England and Wales stabilized or decreased in the 1950s.

The experiences of the three countries indicate that a coronary heart disease pandemic with the same causal factors occurred in all three countries and attained its peak level at various times in the 1950s and 1960s, depending on the age group and the country. It is therefore appropriate to examine the duration of the peak mortality rates and the decreases in mortality rates. The first country to be investigated will be the United States.

References

1. U.S. Bureau of the Census, *Statistical Abstract of the United States: 1970* (Washington, DC: 1970), pp. 806, 810; Dominion Bureau of Statistics, *Canada*

Year Book, 1960: Official Statistical Annual of the Resources, History, Institutions, and Social and Economic Conditions of Canada (n.p.: n.d.), p. 174.

2. Information on Canadian population and vital statistics was laboriously retrieved from Statistics Canada listed on http://www65.statcan.gc.ca/acyb_r000-eng.htm (Accessed 2/27/15)
3. Canada Dominion Bureau of Statistics, *Vital Statistics 1921: First Annual Report* (Ottowa, Canada: Acland, 1923); Canada Dominion Bureau of Statistics, *Vital Statistics 1930: Tenth Annual Report* (Ottowa, Canada: Acland, 1933); Canada Dominion Bureau of Statistics, *Vital Statistics 1935: Fifteenth Annual Report* (Ottowa, Canada: Acland, 1937); Canada Dominion Bureau of Statistics, *Vital Statistics 1940: Twentieth Annual Report* (Ottowa, Canada: Acland, 1942); Canada Dominion Bureau of Statistics, *Vital Statistics 1949 Statistique Vitales*, (n.p.: 1951)
4. T.W. Anderson and W.H. Le Riche, "Ischemic Heart Disease and Sudden Death, 1901-61," *British Journal of Preventive and Social Medicine* 24 (1970):1-9.
5. Canada Dominion Bureau of Statistics, *Vital Statistics Statistique Etat Civil 1952* (Ottowa, Canada: Cloutier, 1954), pp. 60-63.
6. Canada Dominion Bureau of Statistics, *Vital Statistics Statistique Etat Civil 1952*, pp. 60-63.
7. Canada Dominion Bureau of Statistics, *Vital Statistics Statistique Etat Civil 1960* (Ottowa: Duhamel, 1962), pp. 144-50.
8. Jenny R. Podoluk, *Incomes of Canadians* (Ottawa, Canada: Dominion Bureau of Statistics, 1968), p. 159.
9. U.S. Bureau of the Census, *Statistical Abstract of the United States: 1970*, p. 810; John Charlton, "Trends in All-Cause Mortality: 1841-1994," in *The Health of Adult Britain, 1841-1994*, ed. John Charlton and Mike Murphy, 2 vols. (London: Office for National Statistics, 1997), I:25.
10. W.J. Martin, "The Distribution in England and Wales of Mortality from Coronary Disease." *British Medical Journal* June 30, 1956, pp. 1523-25.

Chapter 8

The Peak and Decline of the Coronary Heart Disease Pandemic in the United States, 1970-2010

Coronary heart disease was the leading cause of death in the adult population of the United States at the peak of the pandemic in the 1950s and 1960s. Large differences occurred in the mortality rates of sex, age, and race groups. Mortality rates from the disease began to decrease for all of these population groups in the 1970s and stopped decreasing about 1990 for younger age groups but continued to decrease after 2000 for older age groups. As mortality rates decreased, the differences in mortality rates among the population groups narrowed considerably. These changes demonstrated that the high overall mortality rates and the large differences in the mortality rates of population groups during the peak of the pandemic were produced by the factors that caused the pandemic. In the early twenty-first century post-pandemic coronary heart disease returned to being a disease primarily of the very old, as it had been a century earlier.

To summarize the main decreases in mortality rates from 1970 to 2010, coronary heart disease mortality rates per 1000 at ages 45-54 decreased for white men from 3.3 to 0.8, for white women from 0.7 to 0.2, for black men from 4.2 to 1.0, and for black women from 2.2 to 0.4. At ages 65-74 they decreased for white men from 20.3 to 3.9, for white women from 9.5 to 1.7, for black men from 19.4 to 5.3, and for black women from 13.4 to 2.8.

The Peak Years of the Coronary Heart Disease Pandemic

The 1950s and 1960s were the peak years of the coronary heart disease pandemic in the United States, when all age, sex, and race groups had their highest mortality rates. This peak period was characterized by much higher mortality rates of men than women and of older than younger age groups.

The changes in the International Classification of Diseases categories for coronary heart disease after the emergence of the pandemic had little effect on reported trends in the disease. The symptoms of coronary heart disease were sufficiently distinctive to physicians that they readily adapted to the changes in disease categories. This is indicated by the consistently higher mortality rates of men than women in all age groups and of older than younger age groups of both sexes.

As the coronary heart disease pandemic progressed from its early period about 1940 to its peak about 1970, it became a much more important cause of death for all age groups of adult white and black men and women (Table 8.1). Considering persons ages 35-44 and 45-54, in 1940 coronary heart disease constituted no more than 5 percent of all deaths in all four race-sex groups except for higher rates among white men. In 1970 the

Table 8.1 United States Percentages of All Deaths Caused by Coronary Heart Disease, 1940-2010*

Age Group	*WM*	*WF*	*MO/ BM*	*FO/BF*	*Age Group*	*WM*	*WF*	*MO/ BM*	*FO/BF*
35-44					**55-64**				
1940	9	3	4	3	1940	15	7	5	3
1950	21	6	9	6	1950	35	21	16	13
1960	26	7	12	8	1960	41	26	23	20
1970	24	11	15	13	1970	41	27	29	30
1990	11	8	10	3	1990	26	16	16	16
2010	10	8	6	5	2010	18	10	16	12
45-54					**65-74**				
1940	15	5	5	4	1940	12	8	4	3
1950	33	12	14	11	1950	33	26	16	14
1960	38	14	19	15	1960	39	33	24	23
1970	38	15	24	21	1970	42	38	33	23
1990	22	10	13	11	1990	27	21	19	20
2010	16	7	14	8	2010	17	11	17	14

* 1940: angina pectoris and diseases of the coronary arteries
1950-1960: arteriosclerotic heart disease
1970-2010: ischemic heart disease

WM = white male: WF = white female, MO = male other races (1940-60), FO = female other races (1940-60); BM = black male (1970-2010), BF = black female (1970-2010)

Sources:

1940-60: Robert D. Grove and Alice M. Hetzel, *Vital Statistics Rates in the United States, 1940-1960* (Washington, DC: National Center for Health Statistics, 1968).

1970-2010: CDC Wonder Compressed Mortality File http://wonder.cdc.gov/mortsql.html (Accessed Feb. 28, 2015)

proportion of deaths caused by the disease was at least 11 percent in the four groups ages 35-44 and at least 15 percent for ages 45-54. At ages 55-64 and 65-74 in 1940 the percentage of all deaths caused by coronary heart disease was no more than 8 percent for every race-sex group except for higher rates among white men. In 1970 it was 23 percent or more for every race-sex group and exceeded 40 percent for white men.

Coronary heart disease mortality rates stabilized for age groups 35-44 and 45-54 about 1950 and remained at that level to 1970, with slight differences by race (Tables 4.2 and 8.2). Using arteriosclerotic heart disease mortality rates for 1950 and 1960 and ischemic heart disease mortality rates for 1970, at ages 45-54 the arteriosclerotic heart disease mortality rates per 1000 for white men were 3.2 in 1950, 3.5 in 1960, and the ischemic heart disease mortality rate was 3.3 in 1970. The comparable rates for white women were 0.7 in 1950, 0.6 in 1960, and 0.7 in 1970. Younger black men and women experienced slightly larger increases in their coronary heart disease mortality rates during the 1950-70 period. For example, arteriosclerotic heart disease mortality rates for black men ages 45-54 increased from 2.5 in 1950 to 3.0 in 1960 and the ischemic heart disease mortality rate was 4.2 in 1970. The comparable rates for black women were 1.7 in 1950 and 1960 and 2.2 in 1970.

Among those ages 55-64 in the 1950s and 1960s, coronary heart disease mortality rates were stable for white men and women and increased for their black counterparts (Tables 4.2 and 8.2). Arteriosclerotic heart disease mortality rates per 1000 for white men were 8.1 in 1950, 9.0 in 1960, and the ischemic heart disease mortality rate was 9.0 in 1970. The comparable rates for white women were 2.7 in 1950, 2.5 in 1960, and 2.7 in 1970. Black men and women experienced steady increases in coronary heart disease mortality rates from 1950 to 1970. The arteriosclerotic heart disease mortality rates per 1000 of black men were 5.5 in 1950, 7.2 in 1960, and the ischemic heart disease mortality rate was 9.6 in 1970. The comparable rates for black women were 3.7 in 1950, 4.8 in 1960, and 5.9 in 1970.

At ages 65-74, white and black men and women all experienced increases in coronary heart disease mortality rates from 1950 to 1970, with much larger increases for the black population (Tables 4.2 and 8.2). The arteriosclerotic heart disease mortality rates per 1000 for white men were 16.1 in 1950, 19.1 in 1960, and the ischemic heart disease mortality rate was 20.3 in 1970. Comparable mortality rates per 1000 for white women ages 65-74 were 8.4 in 1950, 9.2 in 1960, and 9.5 in 1970. Among black men ages 65-74 comparable mortality rates increased from 9.4 in 1950 to 13.4 in 1960 and 19.4 in 1970. Among black women the rates increased from 6.6 in 1950 to 9.0 in 1960 and 13.4 in 1970.

During this same period mortality rates from all other causes decreased substantially for both older and younger persons of both races (Tables 4.2 and 8.2). For example, at ages 65-74 from 1950 to 1967 total mortality rates excluding arteriosclerotic heart disease per 1000 decreased for white men from 47.3 to 29.6, for white women from 38.1 to 17.1, for black men from 54.5 to 50.7, and for black women from 44.9 to 34.8.

Age differences in coronary heart disease mortality rates widened from 1950 to 1970 for each of the four race-sex groups. The higher mortality rates per 1000 of those ages 65-74 compared to those 45-54 increased for white men from 12.9 deaths to 17.0 deaths, for white women from 7.7 deaths to 8.8 deaths, for black men from 7.2 deaths to 15.2 deaths, and for black women from 4.9 deaths to 11.2 deaths. Age differences in total mortality rates excluding coronary heart disease decreased over the period.

Widening sex differences in coronary heart disease mortality rates were another feature of the pandemic from 1950 to 1970. Men of each race in every age group had higher mortality rates than women of the same race and their rates increased more than those of women as the pandemic developed, with larger increases for older age groups. The differences between the mortality rates per 1000 of men and women at ages 65-74 increased from 7.7 more deaths for white men than white women in 1950 to 10.8 more deaths in 1970 and from 2.8 more deaths for black men than black women in 1950 to 6.0 more deaths in 1970. At ages 45-54, white men had 2.5 more deaths than white women in 1950 and 2.6 more deaths in 1970 and black men 0.8 more deaths in 1950 and 2.0 deaths in 1970. Sex differences in total mortality rates excluding coronary heart disease decreased for the white population ages 35-44 to 55-64 and increased for the white population ages 65-74 and for all ages of the black population.

Coronary heart disease mortality rates for the black population were lower than for the white population in 1950 but increased more to 1970, especially among older age groups. Among men ages 65-74, coronary heart disease mortality rates per 1000 for white and black men respectively were 16.1 and 9.4 in 1950 and 20.3 and 19.4 in 1970. Among women ages 65-74, the rates for white and black women respectively were 8.4 and 6.6 in 1950 and 9.5 and 13.4 in 1970.

No systematic relationship was found between coronary heart disease mortality rates and total mortality rates excluding coronary heart disease for the four race-sex groups during the 1950-70 period. White men had the highest coronary heart disease mortality rates of the four groups at ages 45-54, 55-64, and 65-74 but black men had the highest total mortality rates excluding coronary heart disease. White women ages 35-44 to 55-64 had both the lowest coronary heart disease mortality rates of the four groups and the lowest total mortality rates excluding coronary heart disease. Black women of the same ages had slightly higher coronary heart disease mortality rates than white women but considerably higher mortality rates from all other causes.

The Decline of the Coronary Heart Disease Pandemic

A key aspect of many pandemics is that the primary victims were population groups that did not have high mortality rates from that disease in normal times. Proof that their high mortality rates were due to the pandemic was that the groups that had the greatest increases in mortality rates during the emergence of the pandemic also had the greatest decreases in mortality rates during its decline. This was shown to be the case for the pandemics of tuberculosis, influenza, and lung cancer in Chapter 1.

This pattern occurred during the coronary heart disease pandemic. Before the pandemic coronary heart disease mortality rates were fairly similar for men and women and for younger and older age groups. As the pandemic emerged, men experienced much greater increases in mortality rates than women at all ages and older age groups experienced much greater increases than younger age groups for both sexes. The decline of the pandemic reversed this pattern and produced greater decreases in the mortality rates of men than women and of older than younger age groups.

The decreases in coronary heart disease mortality rates began in earnest about 1970 for all age and sex groups of white and black men and women using United States vital statistics (Tables 4.2 and 8.2). An important characteristic of the timing of the decrease was that it began at about the same time for all age groups even though the mortality rates of younger age groups had stabilized about 1950 and the rates for older age groups were still increasing or had recently stabilized about 1970. This suggests that novel causal factors came into operation and affected all age groups at the same time to cause the decrease in mortality rates. Mortality rates stopped decreasing about 1990 for younger age groups but continued to decrease to at least 2010 for older age groups.

Trends in mortality rates from 1970 to 2010 can be examined using a single diagnostic category. In 1968 the eighth revision of the International Classification of Diseases introduced a new term, "ischemic heart disease," to replace "arteriosclerotic heart disease," which had been used in the sixth and seventh revisions from 1948 through 1967. The new term was used in that revision (1968-78), the ninth revision (1979-98), and the tenth revision (1999 onwards).[1]

Minor discontinuities in the ischemic heart disease category that occurred among revisions eight, nine, and ten of the International Classification of Diseases had only a small effect on mortality rates. A comparison of ischemic heart disease mortality rates in the years preceding and after each revision indicated that any discontinuities did not have a meaningful impact on the mortality trends for any age group of the four race-sex groups (Table 8.2).

The extent of the decreases in ischemic heart disease mortality rates can be seen using age groups 45-54 and 65-74 (Table 8.2). Ischemic heart disease mortality rates for the four race-sex groups ages 45-54 in 2010

Table 8.2 United States Ischemic Heart Disease and Total Mortality Rates by Age, Sex, and Race, 1970-2010
(Rates per 1,000 persons)

Age Group	*Ischemic Heart Disease*				*Total*			
	WM	*WF*	*BM*	*BF*	*WM*	*WF*	*BM*	*BF*
35-44								
ICD 8								
1970	0.8	0.2	1.4	0.7	3.4	1.9	9.6	5.3
1978	0.6	0.1	1.0	0.4	2.7	1.5	7.1	3.4
ICD 9								
1980	0.5	0.1	0.7	0.2	2.6	1.4	6.9	3.2
1990	0.3	0.1	0.4	0.1	2.7	1.2	7.0	3.0
1998	0.2	0.1	0.3	0.1	2.3	1.2	4.8	2.8
ICD 10								
2000	0.3	0.1	0.4	0.2	2.3	1.3	4.5	2.7
2010	0.2	0.1	0.3	0.1	2.1	1.2	3.1	2.0
45-54								
ICD 8								
1970	3.3	0.7	4.2	2.2	8.8	4.6	17.8	10.4
1978	2.6	0.5	3.5	1.6	7.4	3.9	15.0	7.8
ICD 9								
1980	2.2	0.5	2.6	1.1	7.0	3.7	14.8	7.7
1990	1.2	0.3	1.6	0.7	5.5	3.1	12.6	6.4
1998	0.9	0.2	1.2	0.5	4.8	2.7	10.4	5.7
ICD 10								
2000	1.0	0.3	1.5	0.7	5.0	2.8	10.2	5.9
2010	0.8	0.2	1.0	0.4	4.9	3.0	7.2	4.8
55-64								
ICD 8								
1970	9.0	2.7	9.6	5.9	22.0	10.1	32.6	19.9
1978	6.9	2.1	8.1	4.2	18.0	8.9	28.9	15.6
ICD 9								
1980	5.9	1.8	5.9	3.2	17.3	8.8	28.7	15.6
1990	3.8	1.3	4.3	2.3	14.7	8.2	26.2	14.5
1998	2.6	0.9	3.4	1.7	12.0	7.3	21.8	12.8
ICD 10								
2000	2.8	1.0	4.4	2.1	11.6	7.3	20.8	12.3
2010	1.9	0.6	2.6	1.2	10.3	6.2	16.6	9.7
65-74								
ICD 8								
1970	20.3	9.5	19.4	13.4	48.1	24.7	58.0	38.6
1978	15.9	6.9	15.2	9.6	41.2	20.5	50.8	29.8
ICD 9								
1980	13.8	6.0	11.6	7.3	40.4	20.7	51.3	30.6
1990	9.1	4.0	9.3	5.7	34.0	19.2	49.5	28.7
1998	6.8	3.1	7.3	4.7	30.4	18.9	43.4	27.4
ICD 10								
2000	7.0	3.2	9.1	5.6	29.1	18.7	42.5	26.9
2010	3.9	1.7	5.3	2.8	22.3	15.0	32.1	20.2

WM = white male, WF = white female, BM = black male, BF = black female
ICD 8, 9, 10 = International Classification of Diseases revisions 8, 9, 10
Source: CDC Wonder Compressed Mortality File http://wonder.cdc.gov/mortsql.html (Accessed Feb. 28, 2015)

were about one-fourth of their levels in 1970. Mortality rates for white men ages 45-54 per 1000 decreased from 3.3 in 1970 to 0.8 in 2010, those for white women from 0.7 to 0.2, those for black men from 4.2 to 1.0, and for black women from 2.2 to 0.4. Most race-sex groups ages 65-74 had ischemic heart disease mortality rates in 2010 that were about one-fifth of their levels in 1970. Mortality rates for white men decreased from 20.3 to 3.9, for white women from 9.5 to 1.7, for black men from 19.4 to 5.3, and for black women from 13.4 to 2.8.

The decrease in ischemic heart disease mortality rates produced substantial narrowing of the sex differences in mortality rates in every age group of both races (Table 8.2). The narrowing of the sex differences demonstrated that most of the sex difference was associated with pandemic coronary heart disease, not normal coronary heart disease. At ages 35-44, white men had 0.6 more deaths per 1000 than white women in 1970, which decreased to 0.1 deaths in 2010. For the black population ages 35-44, the male-female differences decreased from 0.7 deaths in 1970 to 0.2 deaths in 2010. At ages 45-54 the male-female difference for whites decreased from 2.6 deaths in 1970 to 0.6 deaths in 2010 and that for blacks from 2.0 deaths to 0.6 deaths. At ages 55-64, the male-female difference for whites decreased from 6.3 deaths in 1970 to 1.3 deaths in 2010 and that for blacks from 3.7 deaths to 1.4 deaths. At ages 65-74 the male-female difference for whites decreased from 10.8 deaths in 1970 to 2.2 deaths in 2010 and that for blacks from 5.0 deaths to 2.5 deaths.

Sex differences in total mortality rates excluding ischemic heart disease also decreased for all groups from 1970 to 2010 (Table 8.2). For example, at ages 45-54 the difference in mortality rates per 1000 from all other causes between white men and women decreased from 1.6 deaths in 1970 to 1.3 deaths in 2010 and at ages 65-74 from 12.6 deaths in 1970 to 5.1 deaths in 2010. For the black population, at ages 45-54 the sex difference in mortality rates between black men and women decreased from 5.4 deaths in 1970 to 1.8 deaths in 2010 and at ages 65-74 from 13.4 deaths in 1970 to 9.4 deaths in 2010.

Age differences in ischemic heart disease mortality rates narrowed during the decline of the pandemic and demonstrated that the large age differences in mortality rates at the peak of the pandemic were a characteristic of pandemic coronary heart disease (Table 8.2). White men ages 65-74 had 17.0 more deaths per 1000 than those ages 45-54 in 1970 and 3.1 more deaths in 2010. White women ages 65-74 had 8.8 more deaths than those ages 45-54 in 1970 and 1.5 more deaths in 2010. Black men ages 65-74 had 15.2 more deaths than those ages 45-54 in 1970 and 4.3 more deaths in 2010. Black women ages 65-74 had 11.2 more deaths than those ages 45-54 in 1970 and 2.4 more deaths in 2010.

The age differences in total mortality rates excluding ischemic heart disease between those ages 65-74 and those ages 45-54 also decreased from 1970 to 2010, but by smaller amounts (Table 8.2). White men ages 65-74 had 22.3 more deaths per 1000 than those ages 45-54 in 1970 and 14.3 more deaths in 2010. White women ages 65-74 had 11.3 more deaths per 1000 than those ages 45-54 in 1970 and 10.5 more deaths in 2010. Black men ages 65-74 had 25.0 more deaths per 1000 than those ages 45-54 in 1970 and 20.6 more deaths in 2010. Black women ages 65-74 had 17.0 more deaths per 1000 than those ages 45-54 in 1970 and 13.0 more deaths in 2010.

The duration of the period of decreases in ischemic heart disease mortality rates varied by age, with decreases occurring from 1970 to 1990 for younger age groups but from 1970 to at least 2010 for older age groups (Table 8.2). Among those ages 45 to 54, the decreases in mortality rates per 1000 persons from 1970 to 1990 varied from 1.5 to 2 deaths for all groups except white women, who had very low mortality rates. The decreases from 1990 to 2010 were small, with no group having decreases of more than 0.6 deaths. Among older age groups, mortality rates continued to decrease until at least 2010. Mortality rates per 1000 for the four race-sex groups ages 65-74 decreased by 5 to 10 deaths from 1970 to 1990 and by 2 to 5 deaths from 1990 to 2010.

The black population experienced a later emergence, peak period, and decline of the pandemic than the white population, particularly among older age groups. In 1950 black men and women ages 55-64 and 65-74 had lower coronary heart disease mortality rate than their white counterparts. In 1970 the mortality rates for black men were very similar to those for white men and black women had higher mortality rates than white women. In the 1990s and 2000s black men and women in age groups 45-54, 55-64, and 65-74 had higher mortality rates than their white counterparts.

Both high and low socioeconomic groups experienced similar decreases in their ischemic heart disease mortality rates during the decline of the pandemic, but higher socioeconomic groups continued to have lower mortality rates throughout the period. This suggests that the socioeconomic differences in mortality rates were caused by factors other than the pandemic. One study of more than two million persons examined two cohorts of persons 45 years of age and older who were recruited by volunteers of the American Cancer Society. The first cohort, recruited in 1959 and followed to 1972, included 445,000 men and 577,000 women. The second cohort, recruited in 1982 and followed to 1996, included 499,000 men and 663,000 women. Education was used to measure socioeconomic level and respondents were divided into five or six categories based on the amount of their education. Age-adjustment was used to standardize the age distributions of each education group.[2]

Trends over time in the coronary heart disease mortality rates of groups with different levels of education can be measured by comparing differences in coronary heart disease mortality rates between the most and least educated groups in the 1959-72 cohort and the 1982-96 cohort. The groups with the most and the least education in each cohort both experienced similar decreases in mortality rates. The data also show a continuation over the period of the higher mortality rates for the least educated groups within each cohort that were found in earlier studies. Coronary heart disease mortality rates per 1000 for the least educated men, with less than nine years of education, were 11.9 for those in the 1959-72 cohort and 6.4 in the 1982-96 cohort, a decrease of 5.5 deaths. The most educated men, who were college graduates, had mortality rates of 9.7 in the earlier cohort and 3.9 in the later cohort, a decrease of 5.8 deaths. The least educated women, with less than nine years of education, had mortality rates of 5.2 in the 1959-72 cohort and 2.9 in the 1982-96 cohort, a decrease of 2.3 deaths. The most educated women, who were college graduates, had rates of 3.6 in the earlier cohort and 1.4 in the later cohort, a decrease of 2.5 deaths. Mortality rates for those with intermediate levels of education were between the two extremes and exhibited the same patterns.

Ischemic Heart Disease Mortality Rates among the Very Old in the Twenty-First Century

The increased proportion of elderly persons in the American population in the twenty-first century warrants an examination of the decline of the pandemic among these groups (Table 8.3). Ischemic heart disease mortality rates started to decrease about 1970 for both younger and older age groups. They stopped decreasing about 1990 for younger age groups but continued to decrease until at least 2014 for older age groups. The group ages 85 and over had the greatest decreases in mortality rates from 2000 to 2014, which is unexpected given the poorer general health of persons of these ages. Among white men ischemic heart disease mortality rates per 1000 decreased from 2000 to 2014 for ages 65-74 from 7.0 to 3.6, for ages 75-84 from 17.4 to 8.9, and for ages 85 and over from 48.1 to 27.5. The comparable decreases for white women for ages 65-74 were from 3.2 to 1.5, for ages 75-84 from 10.4 to 4.8, and for ages 85 and over from 40.3 to 19.5. Similar decreases occurred among the black population.

The narrowing of age differences in ischemic heart disease mortality rates that occurred during the decline of the pandemic also operated for the oldest age groups. The amount of the narrowing was striking considering the short time period involved. Considering age differences, in 2000 white men age 85 and over had 41.1 more deaths per 1000 than white men ages 65-74, which decreased to 23.9 deaths in 2014. The comparable decreases for white women were from 37.1 deaths in 2000 to 18.0 deaths in 2014, for black men from 29.8 deaths to 17.0 deaths, and for black women from 30.5 deaths to 15.5 deaths.

Sex differences in ischemic heart disease mortality rates among the aged decreased by smaller amounts and less consistently from 2000 to 2014. This is to be expected given the small sex differences in mortality rates in 2000. In 2000 among those ages 65-74, white men had 3.8 more deaths per 1000 than white women, which decreased to 2.1 more deaths in 2014. The comparable decreases for black men and women were from 3.5 to 2.3 deaths. Among those ages 75-84, the decreases in sex differences in the white population were from 7.0 to 4.1 deaths and in the black population from 5.6 to 3.7 deaths. At ages 85 and over, sex differences increased slightly from 2000 to 2014 for the white population from 7.8 to 8.0 deaths and for the black population from 2.8 to 3.8 deaths.

Table 8.3 United States Ischemic Heart Disease Mortality Rates, Ages 65 and Over, 2000-2014
(Rates per 1,000 persons)

Age Group	*White Male*	*White Female*	*Black Male*	*Black Female*
65-74				
2000	7.0	3.2	9.1	5.6
2010	3.9	1.7	5.3	2.8
2014	3.6	1.5	4.7	2.4
75-84				
2000	17.4	10.4	18.8	13.2
2010	10.3	5.8	11.5	7.4
2014	8.9	4.8	9.8	6.1
85+				
2000	48.1	40.3	38.9	36.1
2010	31.9	23.8	26.0	22.5
2014	27.5	19.5	21.7	17.9

Source: CDC Wonder Compressed Mortality File http://wonder.cdc.gov/mortsql.html (Accessed Nov. 24, 2016)

Thus the coronary heart disease pandemic peaked in the United States from about 1950 to 1970 and then began a steady decline that was in most respects the opposite of its emergence. Mortality rates decreased for the total population and for all sex, age, and race groups. The differences in mortality rates among age and sex groups that had widened during the rise of the pandemic narrowed during the decline. Mortality rates also decreased for groups with different amounts of education, but the differences between the more and less educated groups did not narrow. Ischemic heart disease in the early twenty-first century became a disease of the very old, which it had been a century earlier.

Experts considered lifestyle risk factors to be the primary causes of the coronary heart disease pandemic. It was found in Chapter 5 that changes in lifestyle risk factors were not associated with the emergence of the pandemic. It is necessary to examine the relationship between changes in lifestyle risk factors and the decline of the pandemic.

References

1. Millicent W. Higgins and Russell V. Luepker, eds., *Trends in Coronary Heart Disease Mortality: The Influence of Medical Care* New York: Oxford University Press, 1988), pp. 279-80.
2. Kyle Steenland, Jane Henley and Michael Thun, "All-Cause and Cause-Specific Death Rates by Educational Status for Two Million People in Two American Cancer Society Cohorts, 1959-1996," *American Journal of Epidemiology* 156 (2002):11-21.

Chapter 9

Explanations for the Decrease in Ischemic Heart Disease Mortality Rates in the United States from 1970 to 2010

This study has demonstrated that a pandemic of coronary heart disease developed in the United States during the 1930s, reached a peak in the 1950s and 1960s, and began to decline about 1970. Most of the decrease in mortality rates resulted from fewer new cases of the disease, demonstrating that the decrease was caused by the decline of the pandemic, not improved treatment of those with the disease. The decrease was not caused by lifestyle changes because no relationship was found between the timing of the decrease and changes in lifestyle risk factors that experts claimed caused the high rates of the disease. No relationship was also found between the timing of the decrease and the use of drugs that reduce high levels of blood cholesterol and blood pressure. These findings are consistent with the findings in Chapter 5 of the absence of a relationship between lifestyle changes and the increase in coronary heart disease mortality rates.

Long-term decreases in ischemic heart disease mortality rates can result from some combination of fewer new cases of the disease and higher survival rates of persons who developed the disease. Fewer new cases of the disease were found to be much more important than higher survival rates in the international MONICA study. The findings of the MONICA study are much more trustworthy than most other studies because of the broader range of social and geographic characteristics of the sample, the long duration of the study, and the sample size that was many times larger than most other studies. In addition, the data gathered were not restricted to mortality. The study examined an extremely large sample of 166,000 fatal and non-fatal coronary heart disease events in men and women ages 35-64 over a lengthy period of more than 10 years in 37 populations in 21 countries on 4 continents, including advanced countries and others. The authors estimated that 79 percent of the decreases in mortality rates for men and 65 percent of the decreases for women were due to fewer new cases. Fewer new cases were even more important in the countries that experienced the greatest decreases in coronary heart disease mortality, which included the United States and other advanced countries.[1]

Experts have attributed the reduction in the number of new cases of the disease to modifications of risk factors in the population by lifestyle changes and drug treatment. The risk factors are the same as those that experts considered responsible for the emergence of the pandemic and described in Chapter 5: cigarette smoking, dietary cholesterol, saturated fats, obesity, hypertension, high blood cholesterol, physical inactivity, and diabetes. Because the coronary heart disease pandemic affected millions of persons and risk factors produce disease in only a small proportion of the persons who experienced them, an extremely large proportion of the total adult population needed to undergo changes in their risk factors to produce the large decreases that occurred in ischemic heart disease mortality rates.[2]

Reduced rates of cigarette smoking, recognized as the single most important preventable cause of morbidity and mortality in all advanced societies, was a minor factor in the decrease in ischemic heart disease mortality rates. Cigarette smoking has a latency period of many years and increases the probability of other chronic diseases in addition to coronary heart disease. Trends in ischemic heart disease mortality rates can therefore be compared to trends in mortality rates from other chronic diseases caused by smoking. If lower rates of cigarette smoking were responsible for the decrease in ischemic heart disease mortality, mortality rates from other chronic diseases caused by cigarette smoking should decrease at appropriate times.

Cancers of the respiratory system are major chronic diseases that are produced in the great majority of cases by the long-term smoking of cigarettes. Mortality rates from cancers of the respiratory system increased from 1970 to 1990 while ischemic heart disease mortality rates were decreasing. For men ages 65-74, deaths per 1000 persons from cancers of the respiratory system increased for white men from 3.4 in 1970 to 4.4 in 1990 and for black men from 3.2 in 1970 to 6.1 in 1990. The rates for women ages 65-74 increased for white women from 0.5 in 1970 to 1.9 in 1990 and for black women from 0.5 in 1970 to 1.7 in 1990. Among men ages 55-64, mortality rates per 1000 white men increased from 2.0 in 1970 to 2.2 in 1990 and for black men from 2.5 in 1970 to 3.8 in 1990. Rates for white women ages 55-64 increased from 0.4 in 1970 to 1.1 in 1990 and for black women from 0.4 in 1970 to 1.2 in 1990. After 1990, the impact of public health programs to reduce smoking steadily reduced mortality rates for cancer of the trachea, bronchus, and lung for practically all of the groups from 1990 to at least 2010.[3]

Thus lung cancer mortality rates began to decrease in older adults about twenty years after coronary heart disease mortality rates began to decrease. It was shown in Chapter 5 that deaths from lung cancer tend to occur at younger ages than deaths from coronary heart disease. Thus it is extremely improbable that decreases in cigarette smoking played a meaningful role in causing the decrease in the coronary heart disease mortality rates.

Components of the diet, especially meats, dairy products, animal fats, and total caloric intake, are considered by many experts to be the primary factors responsible for high rates of coronary heart disease. It was shown in Chapter 5 that these dietary components could not have caused the increase in mortality rates because they did not increase at all or did not increase before the emergence of the pandemic. Consumption of these foods have a cumulative impact, so that most of them would have to decrease substantially at the same times to have a significant impact on ischemic heart disease mortality rates.

Intake of these foods did not decrease in a consistent manner that would make them responsible for the substantial decreases in ischemic heart disease mortality rates that began about 1970. Using federal government data, per person consumption of boneless and trimmed red meats, including beef, veal, lamb, and pork, decreased per year from 132 pounds in 1970 to 126 pounds in 1980 and 112 pounds in 1990. Egg consumption per person decreased from 309 eggs in 1970 to 271 in 1980 and 233 in 1990. Total dairy product consumption per person in milk equivalents was 564 pounds in 1970, 543 pounds in 1980, and 570 pounds in 1990. Consumption of animal and vegetable fats and oils measured in terms of fat content was 53 pounds in 1970, 57 pounds in 1980, and 62 pounds in 1990. In addition, total caloric intake per person increased from a daily average of 3100 calories in 1950-59 to 3200 in 1960-69, 3300 in 1970-79, and 3700 in 1990.[4]

Consumption of specific nutrients considered by experts to be crucial causal factors did not changes in ways that corresponded to the decrease in ischemic heart disease mortality rates. Saturated fat consumption remained unchanged at a daily average of between 59 and 62 grams per person from 1950-59 to 1990. Monounsaturated fat consumption, considered healthier by the same experts, increased from a daily average of 57 grams in 1950-59 to 67 grams in 1990. Dietary cholesterol available for consumption, another important causal factor in the dietary theory, decreased over the period, but mostly after mortality rates began to decrease. Consumption decreased from an average of 510 milligrams per capita per day in 1950-59 to 490 in 1960-69, 460 in 1970-79, 440 in 1980-89, and 410 in 1990.[5]

Thus changes in diet prior to the decline of the pandemic were too small and inconsistent to have produced the substantial decreases in ischemic heart disease mortality rates that occurred after 1970. The inconsistency of the changes in components of the diet is particularly important because the total diet is responsible for the levels of risk. These findings support the evidence in Chapter 5 that dietary components did not change before the emergence of the pandemic in ways that could have increased mortality rates.

Hypertension, blood cholesterol levels, obesity, and diabetes are major risk factors for coronary heart disease. Their prevalence in the population did not change in ways or at times that could have caused them to contribute to the reduction in ischemic heart disease mortality rates. The following age adjusted rates are based on physical examinations of samples of the civilian non-institutional population ages 20-74.

Hypertension rates in the population began to decrease more than a decade after the beginning of the decrease in coronary heart disease mortality rates. Hypertension was defined as a systolic blood pressure of at least 140 mmHg, a diastolic blood pressure of at least 90 mmHg, or taking antihypertensive medication. The proportions of white men with hypertension were 41-45 percent in 1960-62, 1971-74, and 1976-82, and the rate decreased to 26 percent in 1988-94. The comparable rates for white women were 32 percent in the first three periods and 19 percent in 1988-94. The rates for both black men and women were about 50 percent in the first three periods and 35 percent in 1988-94. Another study that examined rates of hypertension (using the same definition) from 1988-91 to 1999-2000 in a sample of the civilian noninstitutional population found that the proportion of the age adjusted sample with hypertension increased from 25 to 27 percent for men and from 25 to 30 percent for women over the period. It also found that in 1999-2000 58 percent of those with hypertension were being treated but that only 53 percent of those being treated had their blood pressure under control. Only 31 percent of all persons with hypertension in 1999-2000 had their blood pressure under control.[6]

The proportion of the population with elevated blood cholesterol levels (defined as at least 240 mg/dL) decreased, but most of the decrease occurred after the beginning of the decrease in coronary heart disease mortality rates. Among white men the proportion with high serum cholesterol was 29 percent in 1960-62, 25 percent in both 1971-74 and 1976-80, and 18 percent in 1988-94. The comparable rates among white women were 35 percent in 1960-62, 28 percent both in 1971-74 and 1976-80, and 20 percent in 1988-94. Among black men the rates were steady at 25 percent in the first three periods and decreased to 16 percent in 1988-94. Among black women the rates were steady at 30 percent in the first two periods, decreased to 25 percent in 1976-80, and to 19 percent in 1988-94.[7]

The percentages of the four race-sex groups who were obese (defined as a body-mass index equal to or greater than 30) remained steady during most of period and increased late in the period. The proportion of white men defined as obese was about 11 percent in 1960-62, 1971-74, and 1976-80, and 20 percent in 1988-94. The comparable rates for white women were about 16 percent in 1960-61, 1971-74, and 1976-80, and 25 percent in 1988-94. Among black men the rates were about 15 percent in the first three periods and 21 percent in 1988-94. Among black women the rates increased from 25 percent in 1960-61 to about 29 percent in 1971-74 and 1976-80, and to 38 percent in 1988-94. These trends are consistent with the increases in per capita total caloric intake described above.[8]

The prevalence of diabetes in the population was stable during the period when coronary heart disease mortality rates decreased. This can be measured by trends in mortality rates. Age adjusted diabetes mellitus mortality rates per 1000 for white men and women were about 0.1 in 1960, 1970, 1980, and 1990 and those for black men and women varied from 0.2 to 0.3 over the period.[9]

These findings demonstrate that changes in the examined risk factors were not responsible for the decline of the coronary heart disease pandemic. Some risk factors did not change in any meaningful way during the period, others changed in ways that are considered undesirable by the experts, others changed years after the onset of the decreases in ischemic heart disease mortality rates, and others did not change for some of the groups that experienced decreases in mortality rates.

The conclusion that risk factors changes did not contribute to the decline of the pandemic is also based on the fact that risk factors have a cumulative impact. Given the very large decrease in mortality rates, many risk factors would have had to change simultaneously in the appropriate directions to make a contribution. Changes in multiple risk factors in the appropriate direction did not occur before the decline of the pandemic.

Prevention programs also include drugs that modify individual risk factors. Two of the most widely used types of drugs to modify risk factors for coronary heart disease in the healthy population (without preexisting heart disease) reduce blood cholesterol and blood pressure levels.

The statin drugs to lower blood cholesterol levels became popular almost thirty years after the beginning of the decrease in ischemic heart disease mortality rates and therefore did not contribute to the decline in the pandemic. Earlier cholesterol lowering drugs had limited use and effectiveness and had no impact on the pandemic. The statin drugs were first approved by the Food and Drug Administration in 1987 and became widely used in the late 1990s. The proportion of a sample of the population in 1988-94 who used a statin drug in the previous thirty days was 4 percent of those ages 45-64 and 6 percent of those ages 65 and over. In 1999-2002 the proportions increased to 14 percent of those 45-64 and 23 percent of those ages 65 and over. In 2007-10 the proportions increased to 22 percent of those ages 45-54 and 47 percent of those ages 65 and over.[10]

Antihypertensive medication became widely used in the 1960s. The proportion of the population who were hypertensive remained unchanged during most of the decrease in coronary heart disease mortality rates, as shown above. Hypertension is also a major risk factor for stroke, but mortality rates from stroke began decreasing several decades before the availability of antihypertensive drugs.[11]

The data presented here have shown that lifestyle changes and the use of drugs designed to modify risk factors for coronary heart disease were not responsible for the decrease in ischemic heart disease mortality rates in the United States. These findings are consistent with the earlier finding that changes in lifestyle risk factors were not responsible for the emergence of the pandemic.

The emergence and peak of the coronary heart disease pandemic in the United States had a substantial geographic component. It increased mortality rates in all states and by different amounts in states with certain characteristics. It produced a group of contiguous states with the highest mortality rates. It is therefore necessary to examine whether the decline of the pandemic reversed these patterns of mortality rates in the states.

References

1. Hugh Tunstall-Pedoe, et al, "Contribution of Trends in Survival and Coronary-Event Rates to Changes in Coronary Heart Disease Mortality: 10-year Results from 37 WHO MONICA Project Populations," *Lancet* 353 (1999):1547-57. See also Earl S. Ford and Simon Capewell, "Proportion of the Decline in Cardiovascular Mortality Disease due to Prevention Versus Treatment: Public health Versus Clinical Care," *Annual Review of Public Health* 32 (2011):5-22; Paul G. McGovern, et al, "Trends in Acute Coronary Heart Disease Mortality, Morbidity, and Medical Care from 1985 through 1997: The Minnesota Heart Study," *Circulation* 104 (2001):19-24; Robert J. Goldberg, et al, "A Two-Decades (1975 to 1995) Long Experience in the Incidence, In-Hospital and Long-Term Case-Fatality Rates of Acute Myocardial Infarction: A Community-Wide Perspective," *Journal of the American College of Cardiology* 33 (1999):1533-39; Ahmet Ergin, et al, "Secular Trends in Cardiovascular Disease Mortality, Incidence, and Case Fatality Rates in Adults in the United States," *American Journal of Medicine* 117 (2004): 219-227.
2. Ford and Capewell, "Proportion of the Decline in Cardiovascular Mortality Disease due to Prevention Versus Treatment," p. 7.
3. National Center for Health Statistics, *Health, United States, 1999* (Hyattsville, MD: 1999), pp. 174-75; National Center for Health Statistics, *Health, United States, 2012* (Hyattsville, MD: 2013), p. 111.
4. U.S. Bureau of the Census, *Statistical Abstract of the United States, 1994* (Washington, DC: 1994), pp. 146-47.
5. U.S. Bureau of the Census, *Statistical Abstract of the United States, 1994,* p. 146.
6. National Center for Health Statistics, *Health, United States, 2002* (Hyattesville, MD, 2002), p. 210; Ihab Hajjar and Theodore A. Kotchen, "Trends in Prevalence, Awareness, Treatment, and Control of Hypertension in the United States, 1988-2000," *JAMA* 290 (2003): 199-206. See also Vicki L. Burt, et al, "Trends in the Prevalence, Awareness, Treatment, and Control of Hypertension in the United States in the Adult US Population: Data from the Health Examination Surveys, 1960 to 1991," *Hypertension* 26 (1995): 60-69.
7. National Center for Health Statistics, *Health, United States, 1999,* p. 222.
8. National Center for Health Statistics, *Health, United States, 1999,* p. 223.
9. National Center for Health Statistics, *Health, United States, 1999,* p. 143.

10. William G. Rothstein, *Public Health and the Risk Factor: A History of an Uneven Medical Revolution* (Rochester, NY: University of Rochester Press, 2003), pp. 297-99; Pradeep K. Nair, Suresh R. Mulukutta, and Oscar C. Marroquin, "Stents and Statins: History, Clinical Outcomes, and Mechanisms," *Expert Review of Cardiovascular Therapy* 8 (2010):1283-95; National Center for Health Statistics, *Health, United States, 2013* (Hyattsville, MD: 2014), p. 292.
11. Theodore A. Kotchen, "Historical Trends and Milestones in Hypertension Research: A Model of the Process of Translational Research," *Hypertension* 58 (2011): 522-38; Douglas J. Lanska and Xiafang Mi, "Decline in US Stroke Mortality in the Era before Antihypertensive Therapy," *Stroke* 24 (1993): 1382-88.

Chapter 10

Decreases in State Ischemic Heart Disease Mortality Rates, 1970-1990

The decline of the ischemic heart disease pandemic after 1970 in the United States reduced age adjusted mortality rates in every state for both men and women. The greatest decreases occurred in the states that had the highest mortality rates at the peak of the pandemic, which demonstrated that the high rates were due to the pandemic. Mortality rates decreased more for men than women in each state and narrowed the sex differences that existed at the peak of the pandemic. The coronary heart disease mortality rates in the states continued to have little correspondence to their mortality rates from all other causes.

The 48 states that were the subject of the 1950 and 1960 analysis in Chapter 6 will be examined from 1970 to 1990 to maintain continuity. The end date of 1990 was chosen because ischemic heart disease mortality rates in younger age groups essentially stopped decreasing after 1990. In addition, state differences in the ischemic heart disease mortality rates of women after 1990 were too small to justify analysis. Only white men and women will continue to be examined because of the heterogeneity and variations in the composition and size of the nonwhite populations in different states. Age-adjusted and sex specific mortality rates are used for each state in every year so that any differences among the states are not due to differences in their age or sex distributions. As in the analysis of the rise of the pandemic, each state is considered an entity and differences in the sizes of the populations of the states are disregarded. Coronary heart disease in 1950 and 1960 was classified as arteriosclerotic heart disease and from 1970 on as ischemic heart disease.

An analysis of the decline of the pandemic in the states of the United States can be better understood by reviewing the mortality patterns in the states during the rise and peak of the pandemic (see Chapter 6). During the peak of the pandemic from 1950 to 1960, every one of the 48 states experienced increases in its age adjusted arteriosclerotic heart disease mortality rates for both men and women. The average arteriosclerotic heart disease mortality rates per 1000 of the 48 states increased from 2.4 in 1950 to 3.0 in 1960 for men and from 1.1 in 1950 to 1.4 in 1960 for women (Table 10.1). The states with lower mortality rates in 1950 experienced greater increases in their mortality rates to 1960, as shown by the decreases in the standard deviations. The mortality rates of men were higher than those of women in all states and increased more over the decade in every one of them. States with higher per capita personal incomes were more likely to have higher age adjusted arteriosclerotic heart disease mortality rates than states with lower incomes.

Table 10.1 United States Coronary Heart Disease and Total Mortality Rates: State Averages and Standard Deviations, 1950-90
(48 state average age adjusted rates per 1,000 persons)

	Arteriosclerotic Heart Disease		*Ischemic Heart Disease*		
	1950	*1960*	*1970*	*1980*	*1990*
State coronary heart disease mortality rates					
White men					
Average	2.4	3.0	3.1	2.1	1.4
Standard deviation	0.41	0.36	0.37	0.26	0.21
White women					
Average	1.1	1.4	1.4	0.9	0.6
Standard deviation	0.27	0.25	0.22	0.14	0.12
State total mortality rates excluding coronary heart disease					
White men					
Average	7.4	6.6	5.9	5.3	5.0
Standard deviation	0.53	0.43	0.47	0.58	0.58
White women					
Average	5.6	4.6	3.5	3.1	3.0
Standard deviation	0.41	0.28	0.23	0.25	0.23

Sources:

1950-60: Robert D. Grove and Alice M. Hetzel, *Vital Statistics Rates in the United States, 1940-1960* (Washington, DC: National Center for Health Statistics, 1968), pp. 663-66, 720-22.

1970-1990: CDC Wonder Compressed Mortality File http://wonder.cdc.gov/mortsql.html (Accessed Feb. 28, 2015)

The increase in mortality rates during the peak of the pandemic was not due to poorer health in the states because their average age adjusted state mortality rates from all other causes decreased from 1950 to 1960. In addition, in both years individual states with higher age adjusted arteriosclerotic heart disease mortality rates did not have higher age adjusted mortality rates from all other causes.

The decade from 1960 to 1970 was a stable period of the peak of the pandemic in the states. No changes occurred during this decade in the average age adjusted state coronary heart disease mortality rates of both men and women (Table 10.1).

Decline of the Coronary Heart Disease Pandemic in the States

The decline of the pandemic in the 48 contiguous states, using the newly adopted term ischemic heart disease, generally reversed the patterns of the rise of the pandemic. Every state experienced decreases in its age adjusted ischemic heart disease mortality rates for both men and women between 1970 and 1990. The states with the highest rates experienced the greatest decreases as shown by the substantial narrowing of the variations in state mortality rates using standard deviations (Table 10.1). This indicates that the large variations in state mortality rates during the peak of the pandemic were due to the pandemic. For white men, the average state age adjusted ischemic heart disease mortality rate per 1000 decreased from 3.1 in 1970 to 2.1 in 1980 and 1.4 in 1990. The standard deviations decreased from 0.37 in 1970 to 0.26 in 1980 and 0.21 in 1990. For white women, the average state mortality rate decreased from 1.4 in 1970 to 0.9 in 1980 and 0.6 in 1990. The standard deviations decreased from 0.22 in 1970 to 0.14 in 1980 and 0.12 in 1990.

Decreases also occurred in the average state age adjusted total mortality rates excluding ischemic heart disease from 1970 to 1990, but the variations in mortality rates among the states increased or remained the same (Table 10.1). Every state experienced decreases in its average age adjusted total state mortality rates excluding ischemic heart disease for both white men and white women during the period. Among white men, the average rates in the 48 states decreased from 5.9 in 1970 to 5.0 in 1990, while the standard deviations increased from 0.47 in 1970 to 0.58 in 1990. Among white women, the average state mortality rates decreased from 3.5 in 1970 to 3.0 in 1990 and the standard deviations remained stable at 0.23 in 1970 and 0.23 in 1990.

The narrowing of the variations in state mortality rates from ischemic heart disease and the stability of the variations in mortality rates from all other causes indicate the operation of different causal factors in the two situations. Variations in ischemic heart disease mortality rates among the states narrowed because of the decline of the national pandemic. The consistency of the state variations in mortality rates from all other causes was due to long term conditions in the individual states, not to national factors.

The decline of the ischemic heart disease pandemic affected men and women in each state in the same way, with the causal factors producing remarkably similar changes in the state mortality rates of each sex. This is indicated by the extremely high correspondence between the state mortality rates of men and women during the peak and decline of the pandemic. At the peak of the pandemic, the correlation coefficients between the state age adjusted coronary heart disease mortality rates of white men and white women were extremely high, about $r = 0.92$ in both 1950 and 1960. As the pandemic declined, this extraordinarily close correspondence continued. The correlation coefficients between the state age adjusted ischemic heart disease mortality rates of white men and white women were $r = 0.92$ in 1970 and $r = 0.88$ in both 1980 and 1990.

State ischemic heart disease mortality rates during the decline of the pandemic continued to have little relationship to state mortality rates from all other causes. States with higher ischemic heart disease mortality rates did not have higher (or lower) mortality rates from all other causes either at the peak or during the decline of the pandemic. For white men the correlation coefficients between state age adjusted ischemic heart disease mortality rates and state age adjusted total mortality rates excluding ischemic heart disease were $r = -0.05$ in 1970, $r = -0.09$ in 1980, and $r = 0.30$ in 1990. For women the coefficients were $r = 0.10$ in 1970, $r = -0.05$ in 1980, and $r = 0.18$ in 1990. These are very similar to the low correlation coefficients for 1950 and 1960 described in Chapter 6.

The decline of the pandemic reduced and then eliminated the positive relationship between state coronary heart disease mortality rates and state per capita incomes that occurred at the peak of the pandemic. As mortality rates decreased, they became more similar among the states for both white men and women, which reduced the relationship between state per capita incomes and mortality rates. This is indicated by the steady decrease in the correlation coefficients between state age adjusted coronary heart disease mortality rates and state per capita incomes. The correlations using arteriosclerotic heart disease mortality rates decreased for white men from $r = 0.65$ in 1950 to $r = 0.40$ in 1960. Using ischemic heart disease mortality rates, the correlations decreased further from $r = 0.23$ in 1970 to $r = -0.09$ in 1980 and $r = -0.18$ in 1990. Using the same terms for white women, the correlation coefficients decreased from $r = 0.61$ in 1950 to $r = 0.48$ in 1960, $r = 0.37$ in 1970, $r = 0.15$ in 1980 and $r = 0.01$ in 1990. State per capita incomes continued to vary widely over the period. The average state per capita income in 1970 was \$3639 with a maximum of \$4871 and a minimum of \$2547. In 1980 the average was \$9392 with a maximum of \$12,170 and a minimum of \$6868. In 1990 the average was \$17,586 with a maximum of \$25,426 and a minimum of \$12,578.[1]

The relationship between state total mortality rates excluding coronary heart disease and state per capita incomes was very weak from 1950 to 1990. The correlation coefficients between the two using total mortality rates excluding arteriosclerotic heart disease for white men were $r = -0.03$ in 1950 and $r = -0.24$ in 1960 and those for women were $r = -0.06$ in 1950 and $r = 0.23$ in 1960. The lack of a consistent pattern continued after 1970 using total mortality rates excluding ischemic heart disease. The correlation coefficients between the two for men were r = –0.46 in 1970, r = –0.24 in 1980, and r = –0.28 in 1990. Those for women were r = –0.03 in 1970, r = 0.17 in 1980, and r = –0.09 in 1990.

One of the most striking characteristic of the coronary heart disease pandemic at its height in 1950 and 1960 was that a contiguous group of states in the northeastern United States had the highest age adjusted mortality rates for men and women (the methods and findings are described in Chapter 6). Evidence that this geographic concentration of high mortality rates was due to the pandemic is shown by the greater geographic dispersion of the states with the highest rates as the pandemic declined. In 1990 eight states had age adjusted ischemic heart disease mortality rates per 1000 for white men of at least 2.4 compared to an all state average of 2.1. Four states (New York, New Jersey, Massachusetts, and Rhode Island) were northeastern states that had been on the 1960 list. Four states had not been on the 1960 list, including one midwestern state (Ohio) and three southeastern states (Kentucky, West Virginia, and North Carolina). The very small variations of the state mortality rates of white women late in the period precluded any analyses.

This analysis of state mortality rates has shown that the decline of the pandemic lowered the mortality rates of every state and reduced the variations in mortality rates among the states. Mortality rates from all other causes also decreased during this period, but the variations among the states remained the same. This indicates that the factors that caused the decrease in pandemic ischemic heart disease mortality rates operated primarily at a national level while the factors that caused the decrease in mortality rates from other causes had an important state component. The decrease in ischemic heart disease mortality rates had the same effect on men and women in each state because their mortality rates corresponded very closely throughout the period. The decrease also eliminated the relationship between coronary heart disease mortality rates and per capita incomes that existed during the peak of the pandemic.

The existence of a geographic component of the pandemic in the United States makes it essential to examine other advanced countries to understand their experiences during the decline of the pandemic.

References

1. Per capita incomes for 1950 and 1960 are taken from U.S. Bureau of the Census, *Statistical Abstract of the United States, 1970* (Washington, DC: 1970), pp. 320-21. Per capita incomes for 1970 are taken from U.S. Bureau of the Census, *Statistical Abstract of the United States, 1980* (Washington, DC: 1980), p. 447. Per capita income for 1980 and 1990 are taken from U.S. Bureau of the Census, *Statistical*

Abstract of the United States, 1994 (Washington, DC: 1994), p. 457. Per capita income for 2000 are taken from U.S. Bureau of the Census, *Statistical Abstract of the United States, 2006* (Washington, DC: 2005), p. 452.

Chapter 11

The Peak and Decline of the Pandemic in Canada, England and Wales, Western Europe, Australia, and New Zealand

The timings and characteristics of the peak and decline of the coronary heart disease pandemic were similar in every one of the seventeen advanced countries on three continents that were studied in this investigation. These many similarities can be explained only by the operation of the same causal factors in every country. Mortality rates peaked and decreased at about the same times in all affected countries. The same population groups in all countries experienced the highest mortality rates at the peak of the pandemic and the greatest decreases in mortality rates as it declined. The geographic regions that experienced the highest mortality rates at the peak of the pandemic experienced the greatest decreases in mortality rates as it declined.

The Peak and Decline of the Pandemic in Canada

The patterns and trends of coronary heart disease mortality rates in Canada from 1950 to 1990 were very similar to those in the United States. At the peak of the pandemic, men had much higher mortality rates than women and older age groups had much higher mortality rates than younger age groups in both countries. The provinces in Canada with the highest mortality rates had characteristics similar to the states in the United States with the highest rates. As the pandemic declined, mortality rates decreased substantially and became more similar among age and sex groups and geographic regions in Canada, as they did in the United States. Minor differences occurred in the intensity and timing of the pandemic in the two countries.

In 1960, during the peak of the pandemic in both countries, arteriosclerotic and degenerative heart disease mortality rates were slightly lower in Canada than in the United States (Table 11.5). Diagnoses were similar in both countries as shown by the higher mortality rates of men than women and older than younger age groups. Age adjusted mortality rates in 1960 per 1000 for men were 3.0 in Canada and 3.3 in the United States and those for women were 1.5 in Canada and 1.6 in the United States. At ages 65-74, the rates for men were 18.3 in Canada and 20.5 in the United States and those for women were 9.4 in Canada and 10.1 in the United States.

Mortality rates began to decrease in Canada in the 1960s for many age and sex groups after having increased from 1950 to 1960 (Tables 7.1 and 11.1). For example, among Canadian men ages 45-49, arteriosclerotic heart disease mortality rates per 1000 increased from 1.9 in 1950 to 2.3 in 1960. Ischemic heart disease mortality rates decreased to 1.9 in 1970 and 0.5 in 1999. Comparable rates for women decreased from 0.5 in 1950 to 0.4 in 1960, 0.3 in 1970, and 0.1 in 1999. Among those age 65-69, arteriosclerotic heart disease mortality rates per 1000 for men were 12.6 in 1951 and 14.9 in 1961, and ischemic heart disease mortality rates were 15.0 in 1970, 11.1 in 1980, and 4.7 in 1999. Comparable rates for women were 6.9 in 1951 and 7.4 in 1961, 6.3 in 1970, 4.4 in 1980, and 1.8 in 1999.

The decreases in ischemic heart disease mortality rates in Canada were greater for men than women and for older than younger age groups (Table 11.1). This narrowed sex and age differences in mortality rates, as occurred in the United States. Considering sex differences in ischemic heart disease deaths per 1000 persons, the higher rates of men than women at ages 45-49 decreased from 1.6 more deaths in 1970 to 0.4 more deaths in 1999, at ages 55-59 from 4.4 more deaths to 1.3 more deaths, and at ages 65-69 from 8.7 more deaths to 2.9 more deaths. Considering age differences, the higher mortality rates per 1000 of those ages 65 than those ages 45-49 decreased for men from 13.1 more deaths in 1970 to 4.2 more deaths in 1999 and for women from 6.0 more deaths to 1.7 more deaths.

The decline of the pandemic produced a narrowing of the large differences in mortality rates among the Canadian provinces that existed in the 1950s, similar to the narrowing of the differences in mortality rates among the states of the United States. In 1952, as shown in Chapter 7, the crude arteriosclerotic heart disease mortality rates per 1000 for Canada were 2.8 for men and 1.7 for women. The highest crude rates in the populous provinces were in British Columbia, with rates of 3.7 for men and 2.1 for women and in Ontario, with rates of 3.6 for men and 2.3 for women. The lowest rates were in Quebec, with rates of 2.1 for men and 1.2 for women. In 1990, mortality rates were much lower and very similar in the three provinces. The national crude ischemic heart disease mortality rates per 1000 decreased to 2.0 for men and 1.5 for women. The rates for British Columbia were now 1.8 for men and 1.3 for women, those for Ontario were now 1.9 for men and 1.6 for women, and those for Quebec were 1.9 for men and 1.4 for women.[1]

Thus the history of the peak and decline of the coronary heart disease pandemic in Canada exhibited patterns that were very similar to those in the United States. These include the levels of peak mortality rates, the duration of peak mortality rates, and the population groups and types of geographic regions that were most severely affected. The decline of the pandemic reduced mortality rates for all groups and regions, with the greatest decreases occurring for those that had the highest rates during the peak of the pandemic. These similarities demonstrate that the rise and fall of the pandemic was produced by the same causal factors in both countries.

Table 11.1 Canada Ischemic Heart Disease Mortality Rates by Age and Sex, 1970-1999
(Rates per 1,000 persons)

Ischemic Heart Disease Mortality					
Age	*Male*	*Female*	*Age*	*Male*	*Female*
45-49			**60-64**		
1970	1.9	0.3	1970	10.0	2.9
1980	1.4	0.3	1980	7.3	2.4
1990	0.8	0.2	1990	4.3	1.4
1999	0.5	0.1	1999	2.8	0.8
50-54			**65-69**		
1970	3.7	0.7	1970	15.0	6.3
1980	2.6	0.5	1980	11.1	4.4
1990	1.4	0.1	1990	7.0	2.6
1999	0.9	0.2	1999	4.7	1.8
55-59					
1970	6.0	1.6			
1980	4.4	1.1			
1990	2.5	0.3			
1999	1.7	0.4			

Sources:
1970: Statistics Canada, *Causes of Death: Provinces by Sex and Canada by Sex and Age 1970* (n.p.: 1971), p. 56
1980: Statistics Canada, *Causes of Death: Provinces by Sex and Canada by Sex and Age 1980* (n.p.: 1982), p. 68
1990: Statistics Canada, *Health Reports: Causes of Death 1990* Supplement No. 11 1992 Vol. 4 No. 1, pp. 72, 225
1999: Statistics Canada, *Causes of Death, 1999: Shelf Tables* (Ottowa, Canada, 2002)

The Peak and Decline of the Pandemic in England and Wales

The peak and decline of the coronary heart disease pandemic in England and Wales was similar in many respects to the experiences in the United

States and Canada. At the peak of the pandemic in all three countries, men had higher mortality rates than women and older age groups had higher mortality rates than younger age groups. As the pandemic declined, the differences between the groups narrowed because men and older age groups each experienced greater decreases in mortality rates than women and younger age groups. The major differences among the countries were that England and Wales had lower mortality rates at the peak of the pandemic and a longer duration of peak mortality rates before the onset of the decline than did the United States and Canada.

During the peak of the pandemic in 1960, arteriosclerotic and degenerative heart disease mortality rates were lower in England and Wales than in the United States and Canada (Table 11.5). Age adjusted mortality rates per 1000 for men were 2.5 in England and Wales, 3.3 in the United States, and 3.0 in Canada, and those for women were 1.2 in England and Wales, 1.6 in the United States, and 1.5 in Canada. At ages 65-74 mortality rates for men were 16.0 in England, 20.5 in the United States, and 18.3 in Canada, and those for women were 7.9 in England and Wales, 10.1 in the United States, and 9.4 in Canada.

The peak of the pandemic in England and Wales occurred from 1951 to 1981 and the start of the decline began after 1981 for most age and sex groups, about a decade later than the start of the decline in the United States (Tables 7.2 and 11.2). This analysis uses arteriosclerotic heart disease mortality rates in 1951 and 1961 and ischemic heart disease mortality rates beginning in 1971. Men in England and Wales ages 65-74 had stable mortality rates per 1000 of 16.5 in 1951, 15.7 in 1961, 16.1 in 1971, and 15.6 in 1981. The rates then decreased sharply to 12.8 in 1991 and 7.5 in 2001. The rates for women ages 65-74 decreased slowly and gradually from 7.9 in 1951 to 6.4 in 1961, 5.6 in 1971, 6.1 in 1981, 5.2 in 1991, and 3.2 in 2001. For men ages 45-54 mortality rates per 1000 increased from 1.7 in 1951 to 2.1 in 1961, 2.7 in 1971, 2.5 in 1981, and then decreased to 1.5 in 1991 and 0.9 in 2001. For women ages 45-54 mortality rates were 0.4 in 1951 and 1961, 0.5 in 1971, 1981, and 1991, and then decreased to 0.2 in 2001. Decreases also occurred for the groups in total mortality rates excluding ischemic heart disease.

Ischemic heart disease deaths in England and Wales increased as a proportion of all deaths from 1971 to 1981 or 1991, depending on age and sex, and then decreased (Table 11.2). Considering men ages 45-54, ischemic heart disease mortality rates as a percentage of total mortality rates increased from about 38 percent in 1971 to about 44 percent in 1981 and then decreased to about 35 percent in 1991 and about 23 percent in 2001. Among men ages 65-74, the proportion increased from about 32 percent in 1971 to about 35 percent in 1981 and about 34 percent in 1991 and then decreased to about 27 percent in 2001. Among women ages 45-54 the proportion increased from about 11 percent in 1971 to about 14 percent in 1981 and about 19 percent in 1991 and then decreased to about 8 percent in 2001. For women ages 65-74 the proportion increased from about 21 percent in 1971 to about 25 percent in 1981 and about 24 percent in 1991 and then decreased to about 18 percent in 2001.

Differences in ischemic heart disease mortality rates between older and younger age groups narrowed in England and Wales as mortality rates decreased (Table 11.2). Comparing ischemic heart disease mortality rates per 1000 persons of those ages 65-74 and 45-54, the differences for men decreased from 13.4 more deaths in 1971 to 13.1 more deaths in 1981, 11.3 more deaths in 1991, and 6.6 more deaths in 2001. The differences for women increased from 5.1 more deaths in 1971 to 5.6 more deaths in 1981 and then decreased to 4.7 more deaths in 1991 and 3.0 more deaths in 2001. Age differences in total mortality rates excluding ischemic heart disease per 1000 also narrowed. For men the differences between the annualized average mortality rates excluding ischemic heart disease for men ages 65-74 and men ages 45-54 narrowed steadily from 30.5 more deaths in 1971-75 to 17.6 more deaths in 2001. Those for women narrowed steadily from 16.9 more deaths in 1971-75 to 11.8 more deaths in 2001.

Sex differences in ischemic heart disease mortality rates in England and Wales decreased during the decline of the pandemic (Table 11.2). For example, at ages 45-54 the number of additional deaths per 1000 among men than women decreased from 2.2 more deaths in 1971 to 0.7 more deaths in 2001. At ages 65-74 men had 10.5 more deaths than women in 1971, which decreased to 4.3 more deaths in 2001. Sex differences in annualized average total mortality rates excluding ischemic heart disease per 1000 narrowed from 1971 to 2001 for older age groups but not for younger age groups. At ages 35-44 and 45-54 no narrowing occurred from 1971-75 to 2001. At ages 55-64, men had 4.8 more deaths than women in 1971-75, which decreased steadily to 1.8 more deaths in 2001. At ages 65-74, men had 14.2 more deaths than women in 1971-75, which decreased steadily to 6.4 more deaths in 2001.

Workers in higher level occupations had lower coronary heart disease mortality rates in a study of English civil servants in the 1970s. The workers were all healthy enough to be employed and had similar health insurance coverage. Their occupational levels were determined by the employing organization and therefore much more internally consistent than census occupational categories. Considering those ages 50-64, 3.2 percent of 6914 administrators and professionals died of coronary heart disease over the period compared to 5.4 percent of 3313 clerical and lower level workers. The lower coronary heart disease mortality rates among higher grade civil service workers remained after considering cigarette smoking rates, blood pressure levels, blood cholesterol levels, and a measure of body mass index.[2]

Another study of socioeconomic group differences in ischemic heart disease mortality rates in England from 1981-83 to 2005-7 found decreases in mortality rates for all socioeconomic groups and a slight narrowing of differences in mortality rates between the higher and lower groups. The study used the Index of Multiple Deprivation 2007, which selected 32,482 small geographic areas, each with about 1500 persons, and divided them into five groups based on education, health, income, crime, and other measures. Each quintile in the analysis included approximately 6.5 million adults ages 35 and over and the same quintiles were used throughout the period.[3]

Table 11.2 England and Wales Ischemic Heart Disease and Total Mortality Rates by Age and Sex, 1971-91 (Rates per 1,000 persons)

Age	*Ischemic Heart Disease Mortality Rate*			*Annualized Average Total Mortality Rate*	
	Male	*Female*		*Male*	*Female*
35-44					
1971	0.7	0.1	1971-75	2.2	1.6
1981	0.5	0.1	1981-85	1.7	1.2
1991	0.3	0.1	1991-94	1.7	1.1
2001	0.2	0.04	2001	1.6	1.0
45-54					
1971	2.7	0.5	1971-75	7.2	4.4
1981	2.5	0.5	1981-85	5.7	3.6
1991	1.5	0.5	1991-94	4.3	2.7
2001	0.9	0.2	2001	3.9	2.6
55-64					
1971	6.9	1.8	1971-75	20.1	10.2
1981	6.9	1.9	1981-85	17.4	9.6
1991	4.9	1.5	1991-94	13.2	7.8
2001	2.7	0.8	2001	10.0	6.3
65-74					
1971	16.1	5.6	1971-75	51.1	26.4
1981	15.6	6.1	1981-85	45.2	24.1
1991	12.8	5.2	1991-94	37.3	21.6
2001	7.5	3.2	2001	28.1	17.4

Sources:

1971-91: John Charlton, "Trends in All-Cause Mortality: 1841-1994," pp. I:23 and John Charlton, et al, "Cardiovascular Diseases," pp. II:65, in *The Health of Adult Britain, 1841-1994*, ed. John Charlton and Mike Murphy, 2 vols. (London: Office of National Statistics, 1997).

2001: Office of National Statistics, *Mortality Statistics Cause: Review of the Registrar General on Deaths by Cause, Sex, and Age in England and Wales, 2001* Series DH-2, no. 28 (London: Office for National Statistics, 2002), pp. 256, 259.

The study found that the annualized average age adjusted ischemic heart disease mortality rates per 1000 in 1981-83 for the least deprived quintile were 5.8 for men and 2.5 for women, for the most deprived quintile were 8.8 for men and 4.1 for women, and for all England were 7.2 for men and 3.2 for women. The comparable annualized average rates in 2005-7 for the least deprived quintile were 2.0 for men and 1.0 for women, for the most deprived quintile 3.9 for men and 1.8 for women, and for all England 2.7 for men and 1.3 for women. Thus over the period the ischemic heart disease mortality rates per 1000 for the most deprived quintile decreased by 4.9 deaths for men and 2.3 deaths for women while that for the least deprived quintile decreased by 3.8 deaths for men and 1.5 deaths for women. This slightly narrowed the differences between the highest and lowest quintiles for men from 3.0 deaths to 1.9 deaths and for women from 1.6 deaths to 0.8 deaths. In both sets of years the rates for the intermediate quintiles were between the two extremes. A study in the United States described in Chapter 8 also found decreases in mortality rates in all socioeconomic groups during the decline of the pandemic. It also found no substantial narrowing of the differences in mortality rates between the higher and lower groups.

Changes in dietary cholesterol and fat intake were unlikely to have been responsible for the decreases in coronary heart disease mortality rates. Weekly per capita consumption of the following food products that contain dietary cholesterol and fat increased or remained constant between 1950 and 1980 for Great Britain: eggs, meats, lard, milk, cheese, and butter.[4]

Thus the experiences of England and Wales during the decline of the coronary heart disease pandemic had many similarities and a few differences to the experiences in the United States and Canada. In all three countries older rather than younger age groups and men rather than women had the highest mortality rates at the peak of the pandemic and experienced the greatest decreases in mortality rates during its decline. A difference was that mortality rates at the peak of the pandemic were higher in the United States and Canada than in England and Wales. Another difference was a longer duration of the peak rates in England and Wales than in the United States and Canada so that mortality rates started to decrease somewhat later in England and Wales.

The Peak and Decline of the Pandemic in Western Europe

The ischemic heart disease pandemic was an important cause of mortality in 13 countries that are considered part of western Europe (including the United Kingdom). The basic characteristics of the pandemic in all countries were similar to those in the United States and Canada, although peak mortality rates were lower in western Europe and the peak rates lasted longer in western Europe. All western European countries experienced decreases in their age adjusted ischemic heart disease mortality rates beginning in the 1980s. Men had higher age adjusted mortality rates than women at the peak of the pandemic in every country and greater decreases in their mortality rates as the pandemic declined. Geographic differences in age adjusted mortality rates also occurred among the countries, with the northern countries having considerably higher mortality rates than the southern countries at the peak of the pandemic. The northern countries experienced greater decreases in their mortality rates as the pandemic declined.

Western Europe in this analysis includes 13 countries based on the generally accepted framework that considers cultures as well as geography (Table 11.3). Germany was excluded because the unification of West and East Germany during the period greatly increased and modified the composition of that country's population. Age adjusted ischemic heart disease mortality rates for each sex in each country were published beginning in 1970 by the World Health Organization Regional Office for Europe. Each country is considered an independent entity and differences in the sizes of their populations are ignored, as in the analyses of the states of the United States and the provinces of Canada. The use of sex specific and age adjusted mortality rates means that any differences among countries cannot be caused by differences in their sex or age distributions. It also precludes analyses of mortality rates in individual age groups.

Table 11.3 Western European Countries Ischemic Heart Disease Mortality Rates by Sex, 1970-2000
(Age adjusted rates per 1,000 persons)

Country	*1970*		*1980*		*1990*		*2000*		*Per Capita 1968 GDP in Dollars ($)*
	M	*F*	*M*	*F*	*M*	*F*	*M*	*F*	
Finland	4.8	1.9	4.3	1.7	3.6	1.6	2.5	1.2	1708
United Kingdom	3.9	1.7	3.8	1.6	3.1	1.5	2.0	0.9	1861
Denmark	3.9	2.1	3.8	1.9	2.9	1.5	1.5	0.8	2545
Ireland	3.8	2.0	3.9	1.8	3.4	1.6	2.3	1.1	1024
Sweden	3.8	2.1	4.1	1.9	2.7	1.2	1.8	0.8	3315
Norway	3.4	1.5	3.1	1.3	2.8	1.2	1.6	0.8	2362
Netherlands	2.9	1.5	2.6	1.3	1.9	1.2	1.3	0.8	1980
Belgium	2.4	1.1	2.0	0.8	1.3	0.5	1.2	0.5*	2154
Italy	1.9	1.1	1.8	0.9	1.3	0.6	1.1	0.5	1418
Switzerland	1.6	0.7	1.8	0.7	1.6	0.7	1.3	0.6	2754
Portugal	1.3**	0.7	1.3	0.7	1.1	0.6	0.9	0.4	529
France	1.1	0.5	1.1	0.5	0.9	0.4	0.8	0.3	2537
Spain	0.8	0.4	1.1	0.5	1.1	0.5	1.0	0.4	773

M = male, F = female, GDP = gross domestic product
*1999, **1971

Source: World Health Organization Regional Office for Europe. http:www.euro.who.int/hfadb (Accessed Feb. 28, 2015)

No analysis was possible to describe the emergence of the pandemic in western European countries because of the lack of appropriate data. However, the experiences of the United Kingdom during the peak and decline of the pandemic were very similar to the other western European countries. This suggests that the detailed analysis of the emergence of the pandemic in England and Wales described in Chapter 7 may be applicable to other western European countries.

In 1960 arteriosclerotic and degenerative heart disease mortality rates in the countries of western Europe were lower than in the United States and Canada as shown by three European countries with among the highest rates. Age adjusted mortality rates per 1000 for men were 2.1 in Denmark, 2.0 in Sweden, and 1.8 in Norway compared to 3.3 in the United States and 3.0 in Canada. The comparable rates for women were 1.2 in Denmark, 1.2 in Sweden, and 0.9 in Norway compared to 1.6 in the United States and 1.5 in Canada. The mortality rates for older age groups were also lower in the three European countries than in the United States and Canada. Among men and women respectively ages 55-65, mortality rates per 1000 were 5.0 and 1.7 in Denmark, 4.9 and 1.5 in Sweden, and 4.8 and 1.4 in Norway, compared to 9.4 and 3.0 in the United States and 8.5 and 2.8 in Canada. Among men and women respectively ages 65-74, mortality rates per 1000 were 13.2 and 7.7 in Denmark, 13.3 and 7.6 in Sweden, and 12.3 and 6.4 in Norway, compared to 20.5 and 10.1 in the United States and 18.3 and 9.4 in Canada.[5]

The ischemic heart disease pandemic was at its peak in the 13 countries of western Europe during the 1970s and then declined (Table 11.3). From 1980 to 2000, every country experienced decreases in its age adjusted mortality rates for both men and women. The average age adjusted mortality rates per 1000 for the 13 countries for men were 2.7 in both 1970 and 1980 and decreased to 2.1 in 1990 and 1.5 in 2000. The comparable rates for women were 1.3 in 1970, 1.2 in 1980, 1.0 in 1990, and 0.7 in 2000.

The countries with the highest age adjusted ischemic heart disease mortality rates in 1970 experienced greater decreases in their ischemic heart disease mortality rates during the decline than countries with lower rates. This is indicated by the narrowing of the variations in the mortality rates among the countries. The standard deviations in age adjusted ischemic heart disease mortality rates for men in the 13 countries decreased substantially from 1.30 in 1970 to 1.21 in 1980, 0.98 in 1990, and 0.53 in 2000. Those for women decreased from 0.62 in 1970 to 0.54 in 1980, 0.47 in 1990, and 0.28 in 2000.

Decreases also occurred in the average age adjusted total mortality rates excluding ischemic heart disease in the 13 countries (Table 11.4). The averages for all countries per 1000 for men decreased steadily from 10.0 in 1970 to 7.2 in 2000. Those for women decreased from 7.0 in 1970 to 4.7 in 2000. The greater decreases for the countries with higher rates are shown by the decrease in the standard deviations. For men the standard deviations decreased from 1.74 in 1970 to 1.61 in 1980, and 0.84 in 2000. For women they increased from 0.98 in 1970 to 1.07 in 1980 and then decreased to 0.54 in 2000.

Table 11.4 Western European Countries Total Mortality Rates by Sex, 1970-2000 (Age adjusted rates per 1,000 persons)

Country	*1970*		*1980*		*1990*		*2000*	
	M	*F*	*M*	*F*	*M*	*F*	*M*	*F*
Finland	16.6	9.7	13.7	7.1	11.9	6.5	9.4	5.3
United Kingdom	14.1	8.7	12.6	7.7	10.6	6.6	8.6	5.7
Denmark	11.5	8.0	11.8	7.2	11.1	7.0	9.2	6.3
Ireland	14.3	10.1	13.6	8.9	11.9	7.3	9.9	6.4
Sweden	10.7	7.3	10.6	6.4	9.1	5.6	7.5	4.9
Norway	11.6	7.3	10.7	6.2	10.1	5.9	8.3	5.2
Netherlands	11.8	7.8	10.8	6.1	10.0	5.7	8.8	5.6
Belgium	14.4	9.3	12.9	7.7	10.6	6.0	9.3*	5.4
Italy	13.0	8.5	11.7	7.0	9.7	5.6	7.9	4.7
Switzerland	12.3	8.0	10.7	6.3	9.4	5.3	7.5	4.6
Portugal	16.8	11.7	14.2	9.1	11.8	7.3	9.8**	5.7
France	12.9	7.4	11.5	6.2	11.4	4.9	8.3	4.4
Spain	11.8	8.2	10.4	6.5	9.6	5.6	8.1	4.5

M = male, F = female
*1999, **1971

Source: World Health Organization Regional Office for Europe data.euro.who.int/hfadb (Accessed Feb. 28, 2015)

The decline of the pandemic after 1980 narrowed the sex differences in the age adjusted ischemic heart disease mortality rates of the 13 countries. Every country had a smaller sex difference in mortality rates in 2000 than in 1980 except Spain. In 1970 in the 13 countries men had an average of 1.4 more deaths per 1000 than women, which increased to 1.5 more deaths in 1980 and then narrowed to 1.1 more deaths in 1990 and 0.8 more deaths in 2000.

Sex differences in average age adjusted total mortality rates from all other causes manifested a different pattern. The sex difference increased in 8 countries from 1970 to 1990 but narrowed in all 13 of the countries from then to 2000. In 1970 the average rate per 1000 for men was 10.1 and that for women was 7.0, a difference of 3.1 deaths. In 1980 the rates for men and women were 9.3 and 5.9, a difference of 3.4 deaths. In 1990 the rates for men and women were 8.4 and 5.1, a difference of 3.3 deaths, and in 2000 they were 7.2 and 4.6, a difference of 2.6 deaths. Data were not available to measure changes in sex differences for individual age groups.

The ischemic heart disease mortality rates of men and women corresponded very closely in the individual countries of western Europe during the decline of the pandemic, as in the states of the United States. This indicates that the causes of the changes in the disease during the decline affected men and women in every country in the same ways. The Pearson correlation coefficients between the age adjusted male and female ischemic heart disease mortality rates in the 13 countries were extremely high: $r = 0.95$ or greater in 1970, 1980, 1990, and 2000. Male and female total mortality rates excluding ischemic heart disease had a very high degree of correspondence in western Europe in 1970 and 1980, but less so in 1990 and 2000. The correlation coefficients were $r = 0.90$ in 1970 and 1980, but dropped to $r = 0.56$ in 1990 and $r = 0.70$ in 2000.

Substantial variations in ischemic heart disease mortality rates occurred among the 13 countries of western Europe at the peak of the pandemic, as in the states of the United States and the provinces of Canada. Male age adjusted ischemic heart disease mortality rates per 1000 in 1970 averaged 2.7 for all countries, ranging from a high of 4.8 in Finland to a low of 0.8 in Spain. Female rates per 1000 averaged 1.3 for all countries and varied from a high of 2.1 in Denmark and Sweden to a low of 0.4 in Spain.

The rankings of the countries based on their age adjusted ischemic heart disease mortality rates remained practically unchanged as the decline of the pandemic reduced mortality rates in every country and narrowed the differences among them. For men the Pearson correlation coefficients for the mortality rates of the countries in 1970 and 1980, in 1980 and 1990, and in 1990 and 2000 were extremely high, all at $r = 0.95$ or greater. For women the same coefficients were equally high, all at $r = 0.94$ or greater. The consistency of the pattern over the entire three decade period is indicated by the very high correlations between the 1970 rankings and the 2000 rankings, with a correlation coefficient of $r = 0.90$ for men and $r = 0.86$ for women. This pattern differed from the United States, where the state rankings were very stable during the peak of the pandemic but fluctuated considerably as the pandemic declined.

Countries with higher ischemic heart disease mortality rates tended to have somewhat lower total mortality rates excluding ischemic heart disease from 1970 to 2000. The relationship weakened at the end of the period, especially for women. Among men the Pearson correlation coefficients between the two mortality rates were $r = -0.42$ in 1970, $r = -0.60$ in 1980, $r = -0.60$ in 1990, and $r = -0.36$ in 2000. Among women, the coefficients were $r = -0.39$ in 1970 and $r = -0.37$ in 1980, $r = -0.05$ in 1990 and $r = 0.18$ in 2000. In the United States during this period practically no relationship was found between state ischemic heart disease mortality rates and state mortality rates from all other causes.

Western European countries with higher ischemic heart disease mortality rates from the peak through the decline of the pandemic tended to have slightly higher per capita gross domestic products (GDP). Per capita GDPs in the 13 countries in 1968 averaged $2151 and varied from a high of $3315 in Sweden to lows of $773 in Spain and $529 in Portugal.

A low positive relationship was found between the 1968 per capita GDPs and the 1970 age adjusted ischemic heart disease mortality rates of the western European countries, with Pearson correlation coefficients of $r = 0.29$ for male mortality rates and $r = 0.31$ for female mortality rates. In 2000 the per capita GDPs adjusted for purchasing power parity averaged $23,600 and varied from a high of $31,400 in Norway to a low of $15,400 in Portugal. The Pearson correlation between the GDPs and the age adjusted ischemic heart disease mortality rates of the countries continued to be low and positive: $r = 0.33$ for men and $r = 0.41$ for women.[6] The United States had a stronger relationship between coronary heart disease mortality rates and per capita incomes at the peak of the pandemic, but the relationship essentially disappeared as the pandemic declined.

The 13 countries of western Europe can be divided into two geographic regions with higher and lower ischemic heart disease mortality rates. In 1970 the eight countries with the highest mortality rates for men and women included every western European country located above the fiftieth parallel: Ireland, United Kingdom, Belgium, Netherlands, Denmark, Norway, Sweden, and Finland. Their age adjusted mortality rates per 1000 varied from that of Finland, with a rate of 4.8 for men and 1.9 for women, to Belgium, with a rate of 2.4 for men and 1.1 for women. All eight countries were geographically contiguous or separated only by bodies of water. The five countries with the lowest ischemic heart disease mortality rates in 1970 were all located below the fiftieth parallel and were also geographically contiguous: Portugal, Spain, France, Switzerland, and Italy. Their mortality rates per 1000 varied from Italy, with a rate of 1.9 for men and 1.1 for women, to Spain, with a rate of 0.8 for men and 0.4 for women.

During the decline of the pandemic, the northern geographic region with higher ischemic heart disease mortality rates experienced greater decreases in its mortality rates than the southern region with lower rates. Between 1970 and 2000 the average mortality rates per 1000 of the eight countries with the highest age adjusted rates in 1970 decreased for men from 2.7 in 1970 to 1.5 in 2000 and for women from 1.3 in 1970 to 0.7 in 2000. The average mortality rates for the five countries with the lowest age adjusted rates in 1970 decreased for men from 1.3 in 1970 to 1.0 in 2000 and those for women from 0.7 in 1970 to 0.4 in 2000. Thus the differences in average mortality rates per 1000 between the two groups of countries narrowed from 1970 to 2000 for men from 1.4 to 0.5 and for women from 0.6 to 0.2. A similar narrowing of differences in mortality rates occurred in the states of the United States and the provinces of Canada as the pandemic declined.

Experts have claimed that the low ischemic heart disease mortality rates of Italy and Spain can be explained by their "Mediterranean diet," which consists of large amounts of grains, fruits, vegetables, and other plant-based foods and small amounts of foods from animals.[7] However, France had age adjusted ischemic heart disease mortality rates from 1970 to

2000 for men and women that were lower than those of Italy and similar to those of Spain. Switzerland had mortality rates that were similar to those of Italy. Neither France nor Switzerland is considered to have diets resembling Mediterranean diets. In 1970 France and Switzerland combined had 38 percent of the total population of the five countries so that their experiences were not aberrations.

Thus the peak and decline of the ischemic heart disease pandemic in the countries of western Europe had many similarities to Canada and the United States. Decreases started to occur in the mortality rates of all of the countries at some time during the decade of the 1970s. As the pandemic declined, the European countries, Canadian provinces, and American states with higher mortality rates at the peak of the pandemic experienced greater decreases in mortality rates than those with lower mortality rates. This narrowed geographic differences in mortality rates in all three locations. Men had higher mortality rates than women at the peak of the pandemic in the United States, Canada, and the countries of western Europe and greater decreases in their mortality rates as the pandemic declined. Older age groups had higher mortality rates at the peak of the pandemic in the United States, Canada, and England and Wales and greater decreases in their mortality than younger age groups as the pandemic declined. The trends in ischemic heart disease mortality rates in all three locations differed in many respects from the trends in their total mortality rates excluding ischemic heart disease.

The causal factors responsible for the pandemic produced an extraordinarily close correspondence between the mortality rates of men and women in every European country and in every American state during the peak and decline of the ischemic heart disease pandemic. During the peak of pandemic, European countries and American states with higher mortality rates for men had higher mortality rates for women. As the pandemic declined and mortality rates decreased, the very close correspondence between the mortality rates of men and women in the European countries and American states continued to be very high.

The ischemic heart disease mortality rates during the peak of the pandemic tended to be somewhat higher in political units with higher scores on some measures of economic wealth. This relationship occurred for the countries of western Europe, the states of the United States, and the provinces of Canada. When the pandemic declined, the relationship no longer existed in the states of the United States and the provinces of Canada but continued in the countries of western Europe.

The major differences between the pandemic in North America and western Europe were its severity and timing. Arteriosclerotic heart disease mortality rates at the peak of the pandemic in the 1950s and 1960s were higher in the United States and Canada than in western Europe. The decreases in ischemic heart disease mortality rates began about 1970 in the United States and Canada but not until about 1980 in western Europe.

The Peak and Decline of the Pandemic in Australia and New Zealand

The coronary heart disease pandemics in Australia and New Zealand were more similar to the United States and Canada than to western Europe. These countries are not analyzed in detail due to the limited availability of detailed historical mortality statistics.

In 1960, Australia had arteriosclerotic and degenerative heart disease mortality rates per 1000 population that were slightly lower than the United States, similar to Canada, and higher than western Europe (Table 11.5). The rates in New Zealand were lower than those in Canada and also higher than western Europe. Age adjusted mortality rates for men per 1000 were 3.0 in Australia and 2.7 in New Zealand compared to 3.3 in the United States and 3.0 in Canada. Those for women were 1.5 in Australia and 1.3 in New Zealand compared to 1.6 in the United States and 1.5 in Canada. Similar patterns occurred for age groups 55-64 and 65-74. Australia and New Zealand had coronary heart disease mortality rates for both sexes that were higher than England and Wales, a northern European country, and considerably higher than Italy, a southern European country.

Table 11.5 Arteriosclerotic and Degenerative Heart Disease Mortality Rates by Age and Sex, Selected Countries, 1960
(Rates per 1,000 persons)

Country	*Age Adjusted*		*Age Groups*			
			55-64		*65-74*	
	Male	*Female*	*Male*	*Female*	*Male*	*Female*
United States	3.3	1.6	9.4	3.0	20.5	10.1
Canada	3.0	1.5	8.5	2.8	18.3	9.4
Australia	3.0	1.5	8.4	3.0	20.0	9.4
New Zealand	2.7	1.3	7.9	2.6	17.0	8.2
England and Wales	2.5	1.2	6.4	1.9	16.0	7.9
Italy	1.6	1.2	3.6	1.6	9.9	7.0

Source: Iwao M. Moriyama, Dean E. Krueger, and Jeremiah Stamler, *Cardiovascular Diseases in the United States* (Cambridge, MA: Harvard University Press, 1971), p. 468.

Ischemic heart disease mortality rates began to decrease in Australia and New Zealand no later than 1980 and continued to decrease in both countries to the early twenty-first century. Age-adjusted ischemic heart disease mortality rates per 1000 in Australia decreased for men from 2.7 in 1980 to 1.9 in 1990, 1.1 in 2000, and 0.7 in 2010. The decreases for women were from 1.3 in 1980 to 1.0 in 1990, 0.6 in 2000, and 0.3 in 2010. The same rates in New Zealand decreased for men from 3.0 in 1980 to 2.2 in 1990 and 1.4 in 2000 and those for women from 1.5 in 1980 to 1.1 in 1990 and

0.7 in 2000. The decreases narrowed the sex differences in deaths per 1000 persons in Australia from 1.4 more deaths for men in 1980 to 0.4 more deaths in 2010 and in New Zealand from 1.5 more deaths for men in 1980 to 0.7 more deaths in 2000.[8]

This analysis of trends in coronary heart disease mortality rates in many advanced countries on three continents provides indisputable evidence that a multinational pandemic caused by the same factors emerged and declined in every one of the countries during much the twentieth century. Considering the emergence of the pandemic, available vital statistics demonstrate that coronary heart disease mortality rates increased rapidly during the 1930s and 1940s in the United States, the province of Ontario in Canada, and England and Wales. Men experienced greater increases in their mortality rates than women and older age groups experienced greater increases in their mortality rates than younger age groups.

The peak and decline of the pandemic were measured for many advanced countries, including the United States, Canada, Australia, New Zealand, and the countries of western Europe. The peak of the pandemic probably began about mid-century in all countries and was present in 1970 in all of them. States of the United States, provinces of Canada, and western European countries with higher per capita incomes tended to experience somewhat higher peak mortality rates than those with lower per capita incomes, although the relationship was not substantial. The decline of the pandemic began at sometime during the 1970s and lowered overall mortality rates with surprising rapidity in all countries. Mortality rates became more similar among population groups because greater decreases occurred for men than women, for older than younger age groups, and for countries and geographic regions with higher mortality rates during the peak of the pandemic.

These findings demonstrate that post-pandemic coronary heart disease differed significantly from the disease during the pandemic. Overall mortality rates were much lower and the differences in the mortality rates of age and sex groups and geographic regions were much smaller. Post-pandemic ischemic heart disease of the twenty-first century has required a reevaluation of the methods of prevention and the risk factors for the disease that were developed during the peak of the pandemic. These issues will be the subject of the concluding chapter.

References

1. These data are taken from the sources in Table 11.1.
2. M.G. Marmot, Geoffrey Rose, M. Shipley, and P.J.S. Hamilton, "Employment Grade and Coronary Heart Disease in British Civil Servants," *Journal of Epidemiology and Community Health* 32 (1978): 244-49.

3. Madhavi Bajekal, et al, "Unequal Trends in Coronary Heart Disease Mortality by Socio-economic Circumstances, England 1982-2006: An Analytical Study," PloS ONE 8(3): March 2013. e59608. doi10.1371/journal.pone.0059608
4. John Charlton and Karen Quaife, "Trends in Diet 1841-1994," in *The Health of Adult Britain, 1841-1994, ed.* John Charlton and Mike Murphy, 2 vols. (London: Office for National Statistics, 1997), pp. I:107-8.
5. Iwao M. Moriyama, Dean E. Krueger, and Jeremiah Stamler, *Cardiovascular Diseases in the United States* (Cambridge, MA: Harvard University Press, 1971), p. 468.
6. Per capita incomes of the countries for 1968: U.S. Bureau of the Census, *Statistical Abstract of the United States, 1970* (Washington, DC: 1970), pp. 320-21. Per capita gross domestic products of the countries for 2000: Eurostat at http:ec.europa.eu (accessed April 11, 2015).
7. Harvey Levenstein, *Fear of Food: A History of Why We Worry About What We Eat* (Chicago: University of Chicago Press, 2012), pp. 132-34, 151-53.
8. Terry Dwyer and Basil S. Hetzel, "A Comparison of Trends in Coronary Heart Disease Mortality in Australia, USA, and England and Wales with Reference to Three Major Risk Factors – Hypertension, Cigarette Smoking and Diet," *International Journal of Epidemiology* 9 (1980): 65-71; K. Uemia and Z. Pisa, "Trends in Cardiovascular Disease Mortality in Industrialized Countries since 1950," *World Health Statistics Quarterly* 41 (1988): 155-78. Age standardized mortality rates after 1980 were obtained from the World Health Organzation at http:who.int/healthinfo/statistics (accessed March 30, 2015). The age standardized rates for Australia and New Zealand are based on the same standard population.

Chapter 12

Coronary Heart Disease after the Pandemic

This analysis has demonstrated that a multinational pandemic of coronary heart disease occurred during the last two-thirds of the twentieth century in many advanced countries, including the United States, Canada, Australia, New Zealand, and the countries of western Europe. The characteristics of the pandemic disease were similar in all countries and strikingly different from those of coronary heart disease early in the century, which was an uncommon disease of the elderly. Mortality rates of all age groups increased rapidly in the 1930s and 1940s and reached peak levels as a major cause of death from about 1950 to the 1970s. Mortality rates during the pandemic were much higher for men than women, for older than younger age groups, and for certain geographic regions. As the pandemic declined beginning in the 1970s, mortality rates in the affected countries decreased by large amounts and decreased more for men, older age groups, and the geographic regions with the highest rates. As a result mortality rates became more similar for men and women, for older and younger age groups, and among geographic regions. In the twenty-first century, ischemic heart disease became primarily a cause of death of the very old, as it had been a century earlier. The different characteristics of ischemic heart disease after the pandemic and the need for restraint in health care expenditures have necessitated reevaluations of methods of prevention of the disease in healthy persons, including accepted risk factors.

The Coronary Heart Disease Pandemic in the Twentieth Century

A coronary heart disease pandemic occurred in most advanced countries throughout the world during the last two-thirds of the twentieth century. It was strikingly different from coronary heart disease early in the twentieth century, which was a minor cause of death with stable mortality rates over time. Before the pandemic most of the victims were elderly and the cause was believed to be hardening of the arteries.

The new pandemic coronary heart disease was similar to many other major pandemics. Relatively rapid increases in mortality rates occurred in large geographic regions with hundreds of millions of residents. This was followed by a period of peak mortality rates in all locations and then a third period when mortality rates decreased relatively rapidly in all locations. These trends differed markedly from trends in total mortality rates excluding coronary heart disease in the same countries during the same period of time.

Another basic characteristic of a pandemic of this type of disease is that its emergence and decline occur at approximately the same times in all geographic areas. This study used available vital statistics to show that increases in coronary heart disease mortality rates began in the 1930s and reached the beginning of a peak period about 1950 in the United States, England and Wales, and Ontario, the most populous province in Canada. Vital statistics are available for 1970 and subsequent years for the United States, Canada, Australia, New Zealand, and the countries of western Europe. They indicate that all of the countries were experiencing the peak period of their mortality rates about 1970. The decline of the pandemic began in all of these countries at some time in the 1970s and continued for a number of decades that varied for different population groups.

A third characteristic of this type of pandemic is that some population groups experience greater increases in their mortality rates as the pandemic emerges and greater decreases in their mortality rates as it declines. Men in all age groups experienced greater increases in their coronary heart disease mortality rates than women as the pandemic emerged and greater decreases as the pandemic declined. Older age groups of both sexes experienced greater increases in their mortality rates than younger age groups as the pandemic emerged and greater decreases as it declined. These trends differed in several respects from the trends of sex and age groups for total mortality rates excluding coronary heart disease.

A fourth characteristic of this type of pandemic is that it produces the highest mortality rates in countries or geographic regions with certain characteristics and that mortality rates vary among the countries or regions. The coronary heart disease pandemic emerged and declined at about the same times in most of the countries with the highest per capita incomes in the world. Variations in mortality rates at the peak of the pandemic occurred among the countries of western Europe, the states of the United States, and the provinces of Canada. The decline of the pandemic substantially reduced mortality rates in every affected country, in every affected location, with greater decreases in countries, states, and provinces with higher mortality rates at the peak of the pandemic.

These characteristics of the coronary heart disease pandemic demonstrate that a single set of causal factors was responsible for the emergence and decline of the pandemic. The causal factors were novel because the characteristics of the victims of pandemic coronary heart disease differed from the victims of normal coronary heart disease early in the century. The factors operated on a worldwide basis because the pandemic emerged and declined at the same times in countries on three widely separated continents.

A striking feature of the pandemic is that the decreases in mortality rates began about the same times in all population groups and geographic locations despite substantial differences in their peak mortality rates. This is contrary to the expectation that the decreases in mortality rates would begin earlier in geographic locations and population groups that experienced milder forms of the pandemic.

The Decline of the Coronary Heart Disease Pandemic

The decline of the coronary heart disease pandemic greatly reduced ischemic heart disease mortality rates in every country and population group that was studied. The countries and population groups that experienced more severe forms of the pandemic had greater decreases in mortality rates, so that mortality rates after the pandemic were more similar among geographic regions, for men and women, and for all age groups. This indicates that the differences in mortality rates among the regions and population groups at the peak of the pandemic were characteristics of pandemic coronary heart disease, not the normal disease.

The lower ischemic heart disease mortality rates and the greater similarities of the rates within and among geographic regions after the pandemic can be demonstrated in the states of the United States and the countries of western Europe. In the 48 states in the continental United States in 1970, the average age adjusted state mortality rate per 1000 white men was 3.1, with the highest state rate of 3.9 and the lowest rate of 2.4, a difference of 1.5 deaths. In 1990, the average decreased to 1.4, with the highest state rate of 1.8 and the lowest 0.9, a difference of 0.9 deaths. In the 13 countries of western Europe the average age adjusted mortality rate per 1000 men in 1970 was 2.7 deaths, with the highest rate of 4.8 and the lowest of 0.8, a difference of 4.0 deaths. In 1990 the average rate decreased to 2.1, with the highest rate of 3.6 and the lowest of 0.9, a difference of 2.7 deaths (Table 11.3), The patterns for women were similar in both regions.

The decline of the pandemic made the ischemic heart disease mortality rates of men and women more similar both within and among countries, as seen in the United States, Canada, and England and Wales. The ischemic heart disease mortality rates per 1000 persons in the United States in 1970 for ages 55-64 were 9.0 for white men and 2.7 for white women, a difference of 6.3 deaths. In 2000 the rates decreased to 2.8 deaths for white men and

1.0 for white women, a difference of 1.8 deaths. The rates in England and Wales for ages 55-64 in 1971 were 6.9 deaths for men and 1.8 deaths for women, a difference of 5.1 deaths. These rates decreased in 2001 to 2.7 deaths for men and 0.8 deaths for women, a difference 1.9 deaths. The rates in 1970 in Canada for ages 55-59 were 6.0 deaths for men and 1.6 deaths for women and for ages 60-64 were 10.0 deaths for men and 2.9 deaths for women, differences of 4.4 and 7.1 deaths respectively. These rates decreased in 1999 at ages 55-59 to 1.7 deaths for men and 0.4 for women, and at ages 60-64 to 2.8 deaths for men and 0.8 for women, differences of 1.3 and 2.0 deaths respectively (Tables 8.2, 11.1, and 11.2).

Age differences in ischemic heart disease mortality rates also decreased both within and among countries as the pandemic declined using the same three countries. In the United States in 1970 white men ages 65-74 had 17.0 more deaths per 1000 than white men ages 45-54. This age difference decreased to 6.0 more deaths in 2000. The comparable decrease in age differences for white women was from 9.0 more deaths in 1970 to 2.9 more deaths in 2000. In England and Wales in 1971 men ages 65-74 had 13.4 more deaths per 1000 than men ages 45-54, which decreased in 2001 to 6.6 more deaths. The age difference for women decreased from 5.1 more deaths in 1971 to 3.0 more deaths in 2001. In Canada in 1970 men ages 65-69 had 13.1 more deaths per 1000 than men ages 45-49, which decreased to 4.2 more deaths in 1999. The same age difference for the two groups of women decreased from 6.0 more deaths in 1970 to 1.7 more deaths in 1999 for the older women (Tables 8.2, 11.1, and 11.2).

These trends provide conclusive evidence that ischemic heart disease after the pandemic differed considerably from the disease during the peak of the pandemic. Mortality rates after the pandemic were much lower and more similar for all geographic regions and age and sex groups both within and among the countries. The geographic regions and age and sex groups with higher mortality rates at the peak of the pandemic experienced greater decreases in their mortality rates. The changes in the characteristics of the disease after the pandemic have great significance for methods of prevention.

These data also demonstrate that the same set of causal factors produced the pandemic in all affected countries. Only a single set of causal factors could produce the same timings of the rise and fall of the pandemic and the same basic changes in the mortality rates of population groups during the pandemic in all countries.

Ischemic Heart Disease after the Pandemic: A Disease of the Elderly

As pandemic coronary heart disease mortality rates decreased in advanced countries at the end of the twentieth century, the pre-pandemic form of the disease, which had affected primarily the elderly, became more prominent. Mortality rates decreased for all age groups during the decline of the pandemic but the oldest age groups had the smallest proportionate decreases. The highest mortality rates after the pandemic were for those ages 75 and over. The older ages of this group than early in the century were due to the increase in life expectancy during the century (Table 12.1).

Ischemic heart disease mortality rates in the United States in 2010 for those ages 55-64 and 65-74 were about one-fifth of the rates in 1970 (Table 12.1). Among those ages 75 and older, mortality rates in 2010 were about one-third of the rates in 1970. As a result, between 1970 and 2010 the percentages of all ischemic heart disease deaths that occurred at ages 75 and over increased for white men from 39 percent to 56 percent, for black men from 26 to 36 percent, for white women from 64 to 79 percent, and for black women from 35 to 60 percent.

Table 12.1 United States Ischemic Heart Disease Mortality by Sex, Age, and Race, 1970 and 2010

Age Group	*Percentage Distribution of Ischemic Heart Disease Deaths*				*Ischemic Heart Disease Mortality Rates per 1000*			
	Male				*Male*			
	White		*Black*		*White*		*Black*	
	1970	***2010***	***1970***	***2010***	***1970***	***2010***	***1970***	***2010***
0-54	12	10	20	18	—	—	—	—
55-64	21	15	24	23	9.0	1.9	9.6	2.6
65-74	28	19	30	23	20.3	3.9	19.4	5.3
75 +	39	56	26	36	50.0	15.7	44.9	14.7
Total	100	100	100	100				
75-84	28	28	19	22	43.2	10.3	33.3	11.5
85+	11	29	7	14	81.6	31.9	47.4	26.0
Age Group	*Female*				*Female*			
	White		*Black*		*White*		*Black*	
	1970	***2010***	***1970***	***2010***	***1970***	***2010***	***1970***	***2010***
0-54	4	4	14	9	—	—	—	—
55-64	9	7	20	14	2.7	0.6	5.9	1.2
65-74	24	11	31	18	9.5	1.7	13.4	2.8
75+	64	79	35	60	37.1	11.8	30.4	11.8
Total	101	101	100	101				
75-84	39	26	22	26	28.9	5.8	26.1	7.4
85+	25	53	13	34	71.9	23.8	44.2	22.5

Source: CDC Wonder at http:wonder.cdc.gov. (Accessed July 23, 2015)

The high proportions of ischemic heart disease deaths at ages 75 and older in the United States in 2010 included many persons who died of the disease at ages 85 and over (Table 12.1). Persons ages 85 and over accounted for 53 percent of all ischemic heart disease deaths among white women, 34 percent of all deaths among black women, 25 percent of all deaths among white men, and 14 percent of all deaths among black men.

It has been suggested that the high ischemic heart disease mortality rates among those ages 75 and over resulted in part from the requirements of United States government death certificates. Practically all very old persons have several comorbidities at death, but one underlying cause of death must be listed on every death certificate. Ischemic heart disease is a convenient and acceptable underlying cause of death.[1]

The inappropriate use of coronary heart disease as an underlying cause of death was found to be of minor importance in an analysis conducted by the original Framingham Heart Study, considered the most rigorous longitudinal study of coronary heart disease in a specific population. The participants in the study were examined every two years by study physicians, who kept detailed records of their health but did not treat them. The participants were treated by their personal physicians. A panel of three study physicians compared the underlying cause of death of the participants, as reported by the study physicians, to the causes listed on death certificates completed by their personal physicians. The sample consisted of 2441 death certificates of study patients from 1948 to 1988 excluding deaths listed as from unknown causes and deaths under age 45. At ages 55-64 the panel considered 182 deaths to be caused by coronary heart disease compared to 178 listed on death certificates. At ages 65-74 the panel considered 266 deaths to be caused by coronary heart disease compared to 283 listed on death certificates. At ages 75 and over the panel considered 262 deaths as caused by coronary heart disease compared to 310 listed on death certificates. Thus the panel agreed with 94 percent or more of the death certificate listings of coronary heart disease as the underlying cause of death in the 55-64 and 65-74 age groups and with 85 percent of the listings in the 75 and older age group.[2]

These findings demonstrate that erroneous death certificate listings of coronary heart disease as the underlying cause of death were an insignificant factor in vital statistics for those under age 75 and a minor factor for those over age 75. This conclusion is strengthened by the study finding that the rate of erroneous reporting did not vary during the forty years of the study, which included the peak years of the pandemic.

The increased concentration of coronary heart disease deaths in the very old requires reevaluations of every accepted risk factor. Risk factors are intended to reduce the probability of disease years or decades after the risk factor develops, but the life expectancies of the very old are short. In 2010 the life expectancy of a man aged 75 years was 11 years and of a woman was 13 years.[3]

Evidence of the need for reevaluation of risk factors in the elderly is found in research on blood cholesterol levels and the benefits of treatment of high levels with statin drugs. A 2016 review of 19 studies with 68,094 elderly participants examined the effect of low density lipoprotein cholesterol levels on total mortality and found no relationship between the two. Major studies of the benefits of reducing blood cholesterol levels with statin drugs in the elderly have varied greatly in their findings.[4] Additional research is required to clarify these findings in the twenty-first century.

Failure of the Coronary Heart Disease Pandemic to Spread to Central and South America

Experts have claimed that the coronary heart disease pandemic in advanced countries resulted from lifestyle changes produced by their high standards of living. If the lifestyle theory is correct, the pandemic should have spread in the late twentieth century to countries with rising standards of living that became more similar to advanced countries. Many countries of Central and South America meet these criteria.

An examination of coronary heart disease mortality rates in Central and South America in the late twentieth century demonstrates that the pandemic did not develop in those countries despite their improved standards of living. Ischemic heart disease mortality rates in Central and South America from 1970 to 2000 were the subject of a study of 10 countries with populations of more than two million persons each and adequate population vital statistics.[5] The accuracy of the diagnoses is supported by the higher ischemic heart disease mortality rates of men than women in every country. Other evidence of the accuracy of the diagnoses is the very close correspondence between the mortality rates of men and women in the 10 countries, with Pearson correlation coefficients of at least $r = 0.9$ in both 1970-72 and 1998-2000. This is similar to the correspondences that occurred in the states of the United States and the countries of western Europe.

The coronary heart disease mortality rates for each sex in the 10 countries were compared between 1970-72 and 1998-2000 with those of the United States and Canada using the same system of age standardization in all 12 countries. In 1970-72 annualized average age adjusted coronary heart disease mortality rates in the countries of Central and South America were much lower than the rates in the United States and Canada. The mortality rates per 1000 in 1970-72 in the United States were 3.2 for men and 1.6 for women and those in Canada were 2.7 for men and 1.2 for women. Argentina had a mortality rate of 1.8 for men and 0.9 for women. None of the following countries had rates greater than 1.4 for men or 1.1 for women: Brazil, Chile, Colombia, Costa Rica, Cuba, Ecuador, Mexico, Puerto Rico, and Venezuela. These were most likely the levels of normal coronary heart disease in the countries.

These Central and South American countries with hundreds of millions of inhabitants did not experience consistent increases in their coronary heart disease mortality rates between 1970-72 and 1998-2000. The mortality rates changed in different ways in individual countries over the period. This is indicated by Pearson correlation coefficients that relate their mortality rates in 1970-72 and 1998-2000, which were $r = 0.4$ for men and $r = 0.6$ for women. Mortality rates for men and/or women increased by small amounts in Mexico, Ecuador, Venezuela, and Costa Rica. They decreased by 1.1 deaths per 1000 for men and 0.6 deaths for women in Argentina and decreased by no more than 0.4 deaths per 1000 for either men or women in the remaining countries.

These vital statistics demonstrate that the coronary heart disease pandemic did not spread to large Central and South American countries late in the twentieth century despite improvements in their standards of living. The coronary heart disease mortality rates of the countries generally remained at their 1970-72 levels. This provides additional evidence that lifestyle changes were not responsible for the rise and decline of the coronary heart disease pandemic in advanced countries.

Ischemic Heart Disease Risk Factors in Healthy Persons in the Twenty-First Century[6]

Ischemic heart disease after the pandemic is a very different disease than it was during the pandemic. One indication is changes in the significance of particular risk factors for the disease in healthy persons. These changes require reevaluations of the risk factors that were identified and measured at the peak of the pandemic. Experts erroneously assumed that the disease would maintain its characteristics indefinitely so that particular risk factors would retain their usefulness.[7]

One proven risk factor that has become less important in preventing ischemic heart disease after the pandemic is obesity in healthy persons. Early in the emergence of the pandemic, obesity was found to be an important risk factor for coronary heart disease and other diseases in healthy persons in methodologically sophisticated studies of hundreds of thousands of policyholders conducted by life insurance companies. Many other rigorous research studies at the peak of the pandemic, including the original Framingham Heart Study, also found evidence of a significant association between obesity and coronary heart disease in healthy persons.[8]

In contrast to this established relationship during the peak of the pandemic, obesity rates in the United States were stable in the 1960s and 1970s and increased by large amounts in the following decades, while ischemic heart disease mortality rates decreased greatly after 1970. The annualized average proportions of age adjusted samples of the population ages 20-74 who were obese (body-mass index of 30 or greater) remained unchanged at 11-13 percent of men and 16-17 percent of women in 1960-62, 1971-74, and 1976-80. The rates then increased to 21 percent of men and 26 percent of women in 1988-94, 28 percent of men and 34 percent of women in 1999-2002, and 35 percent of men and 36 percent of women in 2009-12. The obesity rates from 1976-80 to 2009-12 were very similar for white men, white women, and black men, and substantially higher for black women.[9]

Thus rates of obesity were increasing while ischemic heart disease mortality rates were decreasing. Furthermore, changes in the rates of obesity by sex and race did not correspond to changes in the ischemic heart disease mortality rates of those groups. These contradictory patterns indicated that obesity had become less important as a risk factor for ischemic heart disease after the pandemic. Obesity did not become less important as a risk factor for other diseases, as shown by its continuing importance in the early twenty-first century.[10]

Another change in the importance of a risk factor as the pandemic declined was the greater decreases in the mortality rates of men than women in every age group in every affected country. During the peak of the pandemic male sex was one of the most important risk factors. The decline of the pandemic narrowed sex differences in mortality rates and made male sex less important as a risk factor for ischemic heart disease after the pandemic.

The theory that components of diets and related aspects of lifestyles were major risk factors for coronary heart disease during the pandemic lacks any credibility because of the multinational nature of the pandemic. Substantial increases and decreases in coronary heart disease mortality rates occurred about the same times in all of the countries in North America and Western Europe as well as in Australia and New Zealand. The same population groups in every country experienced the greatest increases and decreases in mortality rates. These similar changes in mortality rates resulted from the operation of the same causal factors in every country. The causal factors could not have included diets and lifestyles, which varied greatly among the countries because of differences in their cultures, climates, geographic locations, economies, societies, and histories. It is beyond the realm of possibility that all of the countries experienced simultaneous identical changes in their diets and lifestyles to cause the emergence of the pandemic and simultaneous identical reverse changes in their diets and lifestyles to cause its decline.

The very large reductions in ischemic heart disease mortality and morbidity rates after the pandemic affect the significance of levels of particular risk factors for healthy persons. The much lower probability of healthy persons developing ischemic heart disease means that the additional risk produced by a specific level of a risk factor has become much smaller than it was at the peak of the pandemic. It is therefore necessary to reevaluate and raise the levels at which modifications of risk factors in healthy persons are likely to be beneficial.

Prevention of Ischemic Heart Disease in the Twenty-First Century

The substantial increases in overall health care expenditures in advanced countries in the late twentieth century have led to the development of new concepts of medical ethics that recognize the responsibilities of health care providers to the entire population. Traditionally, the physician's responsibility was believed to be primarily to the individual patient, so that everything should be done with that patient's best interests in mind. However, substantial increases in health care expenditures in all advanced countries have required changes in this ethical principle. The cost of health care services provided to one patient can result in fewer services provided to other patients, who could be in greater need of them or receive greater benefits from them. Inadequate numbers of health care providers, equipment, and facilities produce a similar problem. The enormous expenditures required to treat and monitor ischemic heart disease risk factors in the healthy population make this issue of great relevance in a period of much lower disease rates after the pandemic.[11]

During the peak period of the coronary heart disease pandemic, a wide variety of extremely expensive programs to identify and modify risk factors were developed for the entire healthy middle-aged and older adult populations.[12] They included frequent periodic medical examinations, frequent medical tests to identify and monitor risk factors, the widespread use of drugs to modify risk factors, mass education and commercial advertising to encourage dietary and other lifestyle changes, and the sale of commercial foods and other products that claimed to reduce the risk of coronary heart disease, These programs expended many billions of dollars annually in the United States and comparable amounts in many other countries. The programs employed many health professionals and created major roles for public and nonprofit organizations and large business corporations in many countries.

This extensive population level approach to prevention in healthy persons continued in the United States long after the substantial reduction in ischemic heart disease mortality rates and changes in the characteristics of persons at greater risk of the disease. One indication is the very large proportion of the population being treated with statin drugs, which require additional expenses for blood tests and physician visits. The statin drugs were the first drugs to lower blood cholesterol levels effectively with fewer adverse effects than earlier drugs. They became available in 1987 and became widely used after 2000 in the United States. Consequently, they were not responsible for the decrease in ischemic heart disease mortality rates, which started about 1970. The annualized average proportions of persons, based on population samples, who took medications to lower their blood cholesterol levels in 1988-94 were 4 percent of both men and women ages 45-64 and 5 percent of men and 6 percent of women ages 65 and over. In 1999-2002 these increased to 17 percent of men and 11 percent of women ages 45-64 and 24 percent of men and 23 percent of women ages 65 and over. In 2007-10 the proportions rose to 25 percent of men and 19 percent of women ages 45-64 and 53 percent of men and 42 percent of women ages 65 and over.[13]

Conclusion

Coronary heart disease during and after the pandemic differed from many other diseases in that enormous expenditures were made not only for treatment of the sick but also for the identification and treatment of risk factors in healthy individuals. The purpose of treating most risk factors is to reduce the probability of the future occurrence of the disease, not to improve current health. Most of the risk factors were identified and their importance measured during the peak of the pandemic.

The much lower rates of post-pandemic ischemic heart disease and changes in the characteristics of healthy persons who develop the disease indicate that recognized risk factors cannot be assumed to have the same significance as they did during the pandemic. Each risk factor must be reevaluated in terms of the new characteristics of the disease.

In addition, during the pandemic judgments were made as to the levels of risk factors that produced sufficient rates of coronary heart disease in healthy persons to warrant treatment, typically erring on the side of greater inclusivity. The much lower rates of ischemic heart disease in the twenty-first century mean that smaller proportions of healthy persons with a given level of a risk factor will develop the disease. Thus the levels of risk factors that benefit from treatment should be raised substantially if the same criteria are applied to post-pandemic ischemic heart disease. The proportion of persons who develop adverse events from the treatments has not changed so that a larger number of persons will experience adverse events with no benefit from the intervention.

The need for a reconsideration of risk factors for ischemic heart disease in healthy persons provides an opportunity to address a fundamental problem of advanced countries in the twenty-first century: the extraordinary large expenditures for health care. During the peak and decline of the pandemic, enormous sums have been expended to modify many coronary heart disease risk factors in healthy persons with the sole objective of reducing their future risk of the disease. The much lower rates of the disease in the twenty-first century permit a substantial reduction in these expenditures with no increased risk to the population. This is the responsibility of health care providers, health researchers, and commercial, public, and non-profit organizations that are concerned with the disease.

References

1. Robert R. Kohn, "Causes of Death in Very Old People," *JAMA* 247 (1982): 2793-97.
2. Donald M. Lloyd-Jones, et al, "Accuracy of Death Certificates for Coding Coronary Heart Disease as the Cause of Death," *Annals of Internal Medicine* 129 (1998): 1020-26.
3. National Center for Health Statistics, *Health, United States, 2012*, (Hyattsville, MD: 2013) p. 76.
4. Timo E. Strandberg, Laura Kolehmainen, and Alpo Vuorio, "Evaluation and Treatment of Older Patients with Hypercholesterolemia: A Clinical Review," *JAMA* 312 (2014): 1136-44; Jerry H. Gurwitz, Alan S. Go, and Stephen P. Fortmann, "Statins for Primary Prevention in Older Adults: Uncertainty and the Need for More Evidence," *JAMA* 316 (2016): 1971-72.; Uffe Ravnskov, et al, "Lack of an Association or an Inverse Association between Low-Desnsity-Lipoprotein Cholesterol and Mortality in the Elderly: A Systematic Review," *BMJ Open* 2016:**6**:e010401.doi:10.1136/bmjopen-2015-010401. Benjamin H. Han, David Sutin, Jeff T. Williamson, et al, "Effect of Statin Treatment vs Usual Care on Primary Cardiovascular Prevention among Older Adults: The ALLHAT-LLT Randomized Clinical Trial," *JAMA Internal Medicine* Published online May 22, 2017. doi:10.1001/jamainternmed.2017.1442.
5. T. Rodriguez, et al, "Trends in Mortality from Coronary Heart and Cerebrovascular Disease in the Americas: 1970-2000," *Heart* 92 (2006) 453-60. The mortality rates for Brazil in 1998-2000 were for several regions of the country.
6. The analyses in this and the following section are based on D.S. Grimes, "An Epidemic of Coronary Heart Disease," *Quarterly Journal of Medicine* 105 (2012): 509-18.
7. Grimes, "An Epidemic of Coronary Heart Disease," pp. 513-14. The continuing emphasis on the same risk factors during and after the pandemic can be seen in *Healthy People 2000, Healthy People 2010*, and *Healthy People 2020* at www.healthypeople.gov.
8. William G. Rothstein, *Public Health and the Risk Factor: A History of an Uneven Medical Revolution* (Rochester, NY: University of Rochester Press, 2003), pp. 64, 214, 283; H.B. Hubert, et al, "Obesity as an Independent Risk Factor for Cardiovascular Disease: A 26-year Follow-Up of Participants in the Framingham Heart Study," *Circulation* 67 (1983): 968-77.
9. National Center for Health Statistics, *Health, United States, 2012*, p. 222.
10. K.M. Flegal, et al, "Cause-specific Excess Deaths Associated with Underweight, Overweight, and Obesity," *JAMA* 298 (2007): 2028-37.

11. Robert D. Truog, "Patients and Doctors – The Evolution of a Relationship," *New England Journal of Medicine* 366 (2012): 581-85.
12. These are described in Rothstein, *Public Health and the Risk Factor*.
13. W. Bruce Fye, *Caring for the Heart: Mayo Clinic and the Rise of Specialization* (New York; Oxford University Press, 2015), pp. 477-78; National Center for Health Statistics, *Health, United States, 2012*, p. 285.

Index